Lecture Notes in Computer S

Edited by G. Goos, J. Hartmanis, and J. van Leeuwen

Springer
Berlin
Heidelberg
New York
Barcelona
Hong Kong
London
Milan
Paris
Tokyo

Caroline Y. Westort   (Ed.)

Digital Earth Moving

First International Symposium, DEM 2001
Manno, Switzerland, September 5-7, 2001
Proceedings

 Springer

Series Editors

Gerhard Goos, Karlsruhe University, Germany
Juris Hartmanis, Cornell University, NY, USA
Jan van Leeuwen, Utrecht University, The Netherlands

Volume Editor

Caroline Y. Westort
University of Applied Sciences of Southern Switzerland (SUPSI)
The CIM Institute for Applied Computer Science and Industrial Technoloy (iCIMSI)
6928 Manno, Switzerland
E-mail: cwestort@cimsi.cim.ch

Cataloging-in-Publication Data applied for

Die Deutsche Bibliothek - CIP-Einheitsaufnahme

Digital earth moving : first international symposium ; proceedings / DEM
2001, Manno, Switzerland, September 5 - 7, 2001. Caroline Y. Westort (ed.).
- Berlin ; Heidelberg ; Barcelona ; Hong Kong ; London ; Milan ;
Paris ; Tokyo : Springer, 2001
    (Lecture notes in computer science ; Vol. 2181)   ISBN 3-540-42586-1

CR Subject Classification (1998): I.3.5, J.6, H.2.8, I.6, J.2

ISSN 0302-9743
ISBN 3-540-42586-1 Springer-Verlag Berlin Heidelberg New York

Springer-Verlag Berlin Heidelberg New York
a member of BertelsmannSpringer Science+Business Media GmbH

http://www.springer.de

© Springer-Verlag Berlin Heidelberg 2001
Printed in Germany

Typesetting: Camera-ready by author, data conversion by PTP-Berlin, StefanSossna
Printed on acid-free paper   SPIN 10840509   06/3142   5 4 3 2 1 0

Preface

Digital manipulation of landform is revolutionizing how our built environment is designed and constructed. On a technical level, three-dimensional geometric modeling of topography has its origins at the interface of geographic information systems (GIS) and computer aided geometric modeling (CAD): the former with its representations of spatial attribute information with digital terrain in several representations (Triangulated Irregular Networks, contour lines, etc.); the latter focusing primarily on the parameterization and combination of geometric primitives. The broadening of these two disciplines to embrace new surveying and navigation advances, e.g. global positioning systems (GPS), together with developments in engineering on the application side, are leading to powerful new suites of functionality. There has been a pronounced need for a forum where these traditionally separate parties can interact.

These proceedings contain the technical papers selected and formally presented as part of the scientific program of the First International Symposium on Digital Earth Moving, 2001 (DEM 2001) held September 5-7, 2001 at the CIM Institute for Computing Science and Industrial Technologies of the University of Applied Science of Southern Switzerland (SUPSI-iCIMSI) in Manno (Lugano), Switzerland. It is the first volume published on this explicit theme. Thirty-six submissions were received, from fifteen countries, with thirteen select papers and posters presented in the official program and in this publication.

DEM 2001 gathered for the first time in one place key representatives from commercial, academic, public, and private sectors who are driving the use and development of these new earth moving capabilities. The disparate domains of civil, petroleum, mining, and construction engineering, landscape architecture and planning, computer science, telecommunications, computational geometry, military, and geomorphology were represented. New commercial software, hardware, research innovations, and application solutions and data requirements were prime topics featured and open for discussion and scholarly debate. All approaches weighed the concerns:

1. Geometric manipulation - editing 3D geometric surfaces,
2. Visualization - seeing a 3D image or model, and
3. Accuracy - correct quantitative information for analysis and construction purposes.

DEM 2001 submission presentations were divided into the following themes:

- Automated Construction
- Topographic Form Parameterization
- "Off-the-Shelf" Software
- Digital Terrain Data
- Real-Time Representation

In addition, a workshop day with the theme, "The Digital Construction Process", was held on September 4 at the Unique Airport Renovation construction site in Zürich. The goal was to provide participants with hands-on access to the tools that currently comprise the new digital earth moving construction cycle – from surveying using GPS, to bulldozer selection and operation, CAD/CAM integration and use, and follow-up quality control. Contractors for the Zürich International Airport Renovation Project have systematized state-of-the-art technology, and participants were treated to direct access to the equipment, together with a detailed project overview.

It was a key objective of DEM 2001 to engage both the technical and the cultural dimensions to new ways of working with digital earth. Traditions within the heretofore industrial cultures involved, must remain open to evolve abreast of innovations if they are to be fruitfully put into action and profited from. The panel discussion held at the conclusion of the symposium laid out an agenda for follow-up meetings and future research and instruction. It is our wish that DEM 2001 will serve as the first of several further international initiatives that bring active parties together in this emerging sub-speciality area. SUPSI-iCIMSI has demonstrated its continuing desire and commitment to facilitate and serve the needs as expressed by participating industrial partners.

The General Co-chairs of the symposium wish to gratefully acknowledge the participation of all of the presenters of papers and posters, the official sponsors of the symposium and their generous support in bringing together key aspects of the technology both to the symposium venue and workshop, and members of the Scientific and Organizing Committees. We also gratefully acknowledge the competent administrative support of Eric Jaminet, Giuseppe Morzanti, and Cinzia Dolci all staff at SUPSI-iCIMSI.

September 2001   Bernardo Ferroni
                                                                 Caroline Y. Westort

Committee

Chairs

Bernardo Ferroni
DEM 2001 General Co-chair
Director, The CIM Institute for Applied Computer Science and Industrial Technology
(iCIMSI) at the University for Applied Sciences of Southern Switzerland (SUPSI).
Manno, Switzerland

Caroline Westort
DEM 2001 Organizer & Scientific Committee Chair
iCIMSI-SUPSI
Manno, Switzerland

Keynotes

Stephen Ervin
Director of Computer Resources
Assistant Dean for Information Technology
Lecturer in Landscape Architecture
Harvard Design School (a.k.a. GSD)
Harvard University, Cambridge, Massachusetts, USA
Coauthor, Landscape Modeling , McGraw-Hill Professional Publishing;
ISBN: 0071357459

Albert Brunner
Project Director
Zürich Airport Extension Project (Zürich Flughafen Aufbau)
Unique Zürich Airport
Zürich Flughafen AG, Zürich International Airport, Switzerland

Advisory Committee
Stephen Ervin, Harvard University, USA
Robert Weibel, University of Zürich, Switzerland

Scientific Committee

Bilal M. Ayyub,University of Maryland, USA
Hansruedi Bär, ETH Zürich, Switzerland
Ian D. Bishop, University of Melbourne, Australia
Kurt Brassel, University of Zürich, Switzerland
Paolo Bürgi, University of Pennsylvania, USA
Jim X. Chen, George Mason University, USA
Leila De Floriani, DISI, University of Genoa, Italy
Stephen Ervin, Harvard University, USA
Bianca Falcidieno, IMA, Genoa, Italy
Michael Goodchild, University of California Santa Barbara, USA

Sponsoring Organizations

The CIM Institute for Applied Computer Science and Industrial Technology (iCIMSI) at the University for Applied Science of Southern Switzerland (SUPSI)

Inter-University Partnership for Earth Observation and Geoinformatics (IPEG).

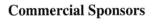 Schweizerische Gesellschaft für Photogrammetri Bildanalyse und Fernerkundung

Geographic Information Systems Division, Department of Geography, University of Zürich (GIUZ).

 HSR
HOCHSCHULE FÜR TECHNIK
RAPPERSWIL

Commercial Sponsors

Pioniere im Tiefbau

SPECTRA PRECISION   Spect a F e.isi.n is no~ cart of ~trnble

autodesk*

CATERPILLAR®

 Geosystems

BENTLEY

 TOPCON

Table of Contents

Zürich Airport Extension Project: Digital Support for Earthwork Construction

Albert Brunner

Project Director
Zürich Airport Extension Project (Zürich Flughafen Aufbau)
Zürich International Airport, Switzerland
Albert.Brunner@uniqueairport.com

Abstract. The expansion of Zurich Airport is one of the major projects to adapt public transport infrastructure in Switzerland to future requirements. Zurich Airport is not among Europe's biggest airports (currently number 7 in terms of movements and number 10 in terms of passengers), but aims at remaining a leading airport in terms of quality, passenger satisfaction and also concerning environmental aspects. Major facility development will help to overcome existing bottlenecks and to cope with future demands. The latest forecasts for Zurich Airport estimate that the volume of traffic handled at the airport will have risen to 34 million passengers and 380'000 movements in 2010. Compared to traffic numbers in 2000, the growth rate is 50 per cent in terms of passengers. Construction activities for the 5 expansion phase take place over almost the entire airport area. And this will take place at an airport already operating beyond its capacity limit. Due to the delays in the legal and concession procedures, realization of the new facilities has become more and more urgent. A very precise phasing-in, including close coordination with the operating requirements, is essential in this case. For a high speed construction process, the contractors use highly sophisticated, modern technical equipment (e.g. GPS, Global Positioning System, for exact excavation or pavement). Main challenge will be to complete this extension within the tight schedule, without interference between construction activities and the airport operation and last but certainly not least, without fatalities.

C.Y. Westort (Ed.): DEM 2001, LNCS 2181, p. 1, 2001.
© Springer-Verlag Berlin Heidelberg 2001

Designed Landforms

Stephen M. Ervin

Director of Computer Resources,
Assistant Dean for Information Technology
Lecturer in Landscape Architecture
Harvard Design School (a.k.a. GSD)
Harvard University, Cambridge, Massachusetts, USA
sme@gsd.harvard.edu

Abstract. Earth is the raw material and foundation for many designed works of architecture, landscape architecture and urban design. From carefully controlled agricultural landforms (terraced rice paddies, contour plowing) to earth-covered structures and structural earth forms, from pre-historic burial mounds to modern 'earthworks art', people have always molded the earth's surfaces to their needs and fancies. Constrained by geological history and climatic imperatives, but also freed by modern technology and structural techniques, designed earthforms take on a wide range of forms. Landscape architects use landforms to control drainage and circulation, to complement designs of buildings and vegetation, and as expressive material in its own right. Techniques of representation of landforms have evolved in the past century from engraving and hachure as epitomized in early topographic maps, to a wide range of technologically sophisticated formats, including laser-gathered point-clouds, triangulated irregular nets, NURB surfaces and grid meshes, cross sections, solid models and procedural and algorithmic descriptions. Modern digital techniques open up new avenues for describing and building earthforms - but still to serve our needs and fancies.

C.Y. Westort (Ed.): DEM 2001, LNCS 2181, p. 2, 2001.
© Springer-Verlag Berlin Heidelberg 2001

Terraffectors and the Art of Consensus Building

Gary Huber

President & Lead Feature Programmer
3D Nature LLC

Abstract. We are an earth-moving species. The surface of the earth the world over is testament to that fact. From the irrigation canals and funerary mounds of ancient cultures to modern day dams, mines, quarries, road cuts, basements and golf courses the evidence is plain: We are obsessed with reshaping the world beneath our feet. Earth sculpting has always been and remains an expensive endeavor. The carpenter's adage, "Measure twice, cut once," applies here as well. In today's competitive markets, thorough planning and accurate volume assessment are more important than ever. Digital technologies are emerging as critical components of the planning process. Not only do they expedite the tedious drafting chores involved with any significant construction project but, coupled with high-resolution terrain models and aerial imagery, they allow the engineer far greater insight and computational accuracy than was previously possible. Financial and social concerns frequently converge upon the design engineer encouraging the evaluation of multiple design alternatives and creative solutions. The final design must satisfy a wide range of stakeholders' objectives including maximum aesthetic appeal, minimal environmental damage, and of course financial viability. Again digital technologies have much to offer when it comes to optimizing design among such disparate constraints. Communication of ideas between designer and stakeholders is the key to resolving differences and visualization is the key to communication. Software especially enabled for visualizing the natural and engineered landscape is a valuable asset to designers who need to communicate. Software with these capabilities is commercially available and has been helping stakeholders reach consensus for nearly a decade. In particular one software package, World Construction Setâ, embodies a unique and powerful approach to modeling engineered elements of the landscape. It is a procedural landform geometry engine built into a photorealistic renderer. Using terrain models, lines, polygons, cross-section profiles, material descriptions, foliage images and 3D objects, the esoteric plans of the designer are transformed into understandable, believable representations - pictures which accurately and artistically communicate the designer's ideas to the technical and non-technical audience alike. These digitial tools have been dubbed Terraffectorsä and have evolved to fill the visualization needs of transportation, mining, reservoir and site planning engineers, golf course and ski resort designers, landscape architects and even the film and game development industries, all of whom have a common need to portray their unique view of the earth. As Terraffector technology is adapted by this wide spectrum of users new needs are exposed. The technology continues to evolve in an ongoing effort to fulfill these needs. The future of this technology is as much in the hands and minds of its users as it is of its developers. Recent and imminent Terraffector advances are illustrated in this presentation along with a discussion of trends that guide this Terraffected path we travel.

C.Y. Westort (Ed.): DEM 2001, LNCS 2181, p. 3, 2001.
© Springer-Verlag Berlin Heidelberg 2001

GPS –Based Earthmoving for Construction

Craig L. Koehrsen, William C. Sahm, and Claude W. Keefer

Caterpillar Inc., 100 N. E. Adams AB9740, Peoria, Illinois, 61629, USA
{koehrcl, sahmwc, keefecw}@cat.com

Abstract. Advances in on-board machine computing, display, and wireless communication technology have revolutionized earthmoving in construction. Efficient construction operation demands that the right people have the right information at the right time.   Caterpillar's Computer Aided Earthmoving System (CAES) meets this demand by providing real-time integration of planning and operations by providing the means to display plans to the machine operator immediately after they are created. CAES also turns the earthmoving machine into a survey instrument, which is used to continuously survey the terrain as the machine works. In addition to providing terrain data, other information, such as productivity and machine utilization, is available from the Caterpillar Computer Aided Earthmoving System.

1   Introduction

Caterpillar's Computer Aided Earthmoving System (CAES), which was developed in conjunction with Trimble Navigation, is used on construction machines to give operators information about earthmoving tasks to be performed in a concise and easily understood manner. CAES allows the machine operator to view the machine's location and a color-coded work plan in real-time. As the operator works, terrain data and machine position information are updated using Real-Time Kinematic (RTK) Global Positioning System (GPS) technology providing continuous information to the operator about his progress toward completion of a design. With its on-board display and electronically transmitted data, CAES dramatically improves the efficiency of earthmoving production. CAES also greatly enhances site operations by providing site management personnel with information on the status of the work and the productivity of the machines performing the work.

2   Typical Machine Applications and System Components

CAES is the software that runs on the computer on-board the earthmoving machine. CAES combines GPS position data with information received from the construction office via a radio network to provide a daylight readable color display to the operator. Using this display, the operator is able to work to a design without relying on survey stakes and flags.

   CAES can be used on most earthmoving equipment with the most widely used machines including track-type tractors, hydraulic front shovels, cable or rope shovels, compactors, motor graders, wheel tractor scrapers, and wheel loaders. CAES applications include grade material, material control, and compaction. This document will concentrate on grade control applications.

C.Y. Westort (Ed.): DEM 2001, LNCS 2181, pp. 4–17, 2001.

In addition to the on-board system, there is a suite of office products and field hardware. The basic system components associated with all of these applications are identified in the following table.

Table 1. Basic System Component

Component	Description
METS Office Software	METScomms and METSmanager make up the METS office software and are used to manage file conversions and site terrain data manipulation. METScomms reads design files in standard DXF (AutoCAD) formats and converts them to CAES (CAT) format files. METS then sends designs to machines over the radio network and keeps the master site models up-to-date by merging terrain data sent from machines. It also schedules productivity and diagnostic reports from machines.
CAESoffice	CAESoffice gives the user the ability to view design files before they are sent to the machines and to compare these designs to the current state of the construction site. CAESoffice also allows the user to view machine locations and updated terrain models from CAES-equipped machines in near real time.
Productivity Data Viewer	Reporting tool used to display productivity information sent from the machines. Productivity information includes volumes of material moved, material types and machine utilization.
Global Positioning System Reference Station	Continuously monitors GPS data in timing, measurement, etc., then transmits or records reference information for those errors. The GPS reference station helps in obtaining the most accurate positioning data possible.
Radio Network	Transmits GPS reference data to all machines and routes design files, diagnostics, productivity data and terrain updates to appropriate location.
CAES On-Board Software and Hardware	CAES Software Rugged High-Resolution Color Display and Computer (CD1040) TC900 Dual Port Radio GPS 740 Receiver and Power Supply L1/L2 GPS Antenna

3   Design Process

Caterpillar provides construction planners or surveyors with the ability to create a design, evaluate the design while comparing it to the current state of the construction site, make changes or refinements, and deliver it to a machine operator without creating a paper plot or driving a survey stake. The process can all be accomplished from the construction office with two-way communication between the office and the earthmoving machines.

3.1   The Earthmoving Process

Construction design generally includes using a construction planning or survey software package to create an elevation design to which the machine operators should work. This design may be a simple production dozing design such as a dual sloped plan for field drainage or a much more complex design such as super-elevated curves.

Designs are developed using a survey of the site and the identification of various natural and man-made structures. Traditionally, these designs are transferred to the field by placing survey stakes at key locations. The number of stakes that must be used depends on the complexity of the design. It is impossible to supply enough survey stakes to convey the same resolution that was in the original design. Field foremen and paper plans usually supplement the survey information. As the job progresses, additional surveys may be required to convey information on the next phase, to make changes to the current work, and to document completed work against plan. This conventional process is time-consuming and it contains numerous opportunities for error.

Designs are developed using familiar software packages and then exported into a standard DXF file. Using CAES, the design may now be easily transferred to the operator by sending it over the radio network to the on-board display. This process quickly provides the operator with the entire design, which can be used, to better plan the work, while maintaining the original design resolution. The need for staking is eliminated and the as-built surfaces are returned to the office in near real-time. Changes in the design, which would traditionally require re-staking, are quickly completed by sending a new design file to the machine.

The following sections describe in detail the design process using CAES in the earthmoving information age.

3.2   CAES Data Flow

Once a project has been selected, the first step is typically to obtain the existing, or original, ground topography. The design engineer creates a plan using one of a multitude of civil engineering software design packages. The designs are exported from the planning software into a standard format, such as AutoCAD DXF, and then passed on to an office software application called METS. METS consists of METScomms, which does the actual file conversion and communications, and METSmanager, which provides the user interface for METScomms. The DXF file is converted to a CAT file format, which is compatible with the CAES on-board software. The converted file may then be viewed with CAESoffice to verify that it depicts the desired plan and is compared to the existing topography. The design is then sent across the radio network to CAES on-board the machine.

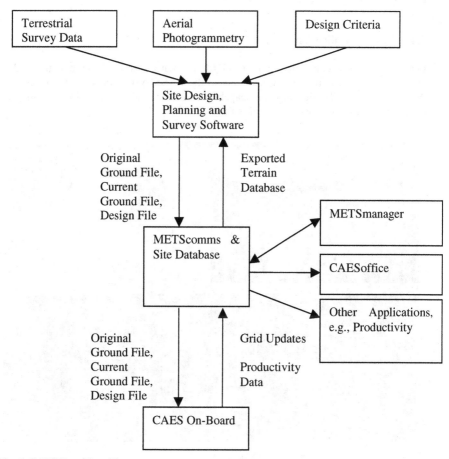

Fig. 1. CAES Data Flow Diagram

The CAES Data Flow diagram depicts how data from original sources is transferred from the survey software to METS and then transferred out to the CAES-equipped machine. As the terrain is updated in the field, CAES sends pieces of terrain data back to METScomms, which are used to update the site terrain data. The data may then be exported back to the design planning software. CAES also sends productivity information on volumes of material moved and machine utilization.

3.3   Design Conversion

The elevation design work for the site is done using the construction planning or survey software package. Preferably, elevation designs are represented as a Triangular Irregular Network (TIN) comprised of 3DFACES or closed 3DPOLYLINEs. A CAES compatible format file is then created from the TIN that is provided. Alternately, the elevation design can be represented as a set of POINTs,

LINEs, and 3DPOLYLINEs that accurately describe the design surface. METScomms will then convert the design to a CAES compatible TIN file that will be used by the on-board system.

3.4   Design Verification

Once the design is completed and a CAT file is created, the CAT file containing the elevation design can be sent immediately to the CAES system onboard the machine. Additionally, CAESoffice provides capabilities to preview the elevation design and compare it to the existing topography of the site.

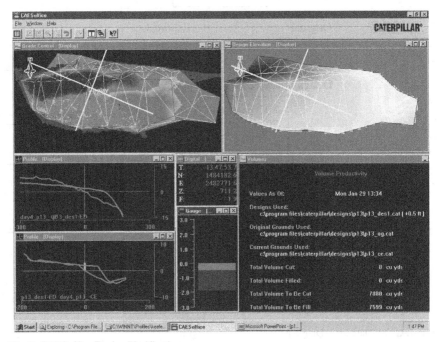

Fig. 2. CAESoffice Design Verification

Figure 2 shows a CAESoffice session used to verify a simple dozing design. Plan views, with or without displaying TIN lines, can be used to verify cuts and fills, while specific elevations of the design may be examined using a virtual mouse-driven machine.

The upper left plan view shows the cuts and fills on the design versus the current ground. In this session, areas to be cut are shown in red (medium gray), areas that are on-grade are green (light gray), and areas to be filled are blue (dark gray). The upper right view is an elevation picture of the design. The display shows how varying shades of gray are used to display the different elevations. Zooming functions allow users to look at specific design areas in more detail. The TIN line display is especially useful for determining if the construction planning or survey software has created a TIN that accurately reflects the intentions of the designer.

The two profile views show how the design and current ground varies. Profiles may be displayed parallel to or perpendicular to the direction of the machine. Two other windows also give views of the point of interest on the display. The digital view displays the virtual machine position. This window provides the CAESoffice user with positions in site coordinates so spot elevation checks can be done for any point on the design. The gauge view graphically shows the cut or fills at a given point.

The productivity window in the lower right corner displays the values to cut and fill. These values give a quick check of how much material must be added or removed to complete the design.

These CAESoffice features give the user the ability to quickly evaluate multiple elevation designs in great detail without leaving the office. Design revisions can be viewed and corrections and improvements made before the designs are implemented in the field.

3.5   Design Transfer

After verifying that the CAT elevation design file is accurate, the user can deliver the design to the machine operator. The user selects the CAT file in the file window and drags it to the machine icon. This action prompts the user to specify a priority for the file transfer and indicate whether this file should be displayed immediately to the operator or stored onboard to be selected at a later time. Ideally, designs are created, converted, and verified before they are needed on the machine.

3.6   Data Collection

CAES provides construction personnel with data that has historically been very difficult or even impossible to obtain. Examples of this type of data are continuous surveys of each area in which a machine is working and up-to-the-minute terrain maps showing progress toward the final design throughout the construction process. CAES also sends productivity and diagnostic information to the office for further processing. Wireless transfers of designs from the office to the field may be sent as soon as changes are identified. Because Caterpillar transfers this type of information throughout the construction site over the wireless network and the existing construction computer network, construction personnel can have the information they need when they need it.

4   On-Board CAES Display

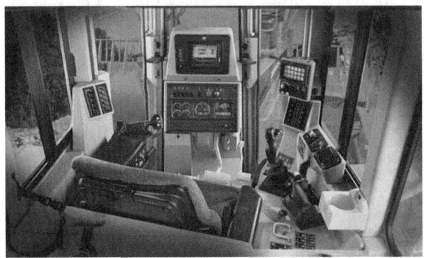

Fig. 3. CAES Display in Track-Type Tractor Cab

CAES was developed by Caterpillar to provide machine operators with the information they need to do their job more efficiently.   It does this by giving the operator a real-time view of required cuts and fills based on the current elevation and the design. This process effectively combines planning and field operations. The CAES display is a rugged easy to use system that incorporates RTK GPS data with design information.

Unlike grade stakes or other external visual design aids, the CAES display may be used day or night with the same consistent results. This grade control display can effectively replace surveyed stakes if desired, or it can complement existing methods of earthmoving by reducing surveying cycles and improving operator efficiency.

4.1   Display Configuration

CAES displays are highly configurable. An office system is used to select the windows desired for the particular operation. Multiple screens may be configured each having multiple windows. Window selections include plan views, profile views, gauges, and a variety of text based informational windows. The following sections describe some of the available windows and show a sample configuration. Once the window configuration is complete, it is sent to the machine where the desired windows are displayed.

The tool bar at the top of the screen is also configurable. The order and type of keys are selected when the windows are configured. Individual tools are selected from a palette of available functions. Each screen may have its own set of tools selected for the specific application. During normal operation, very little interaction is needed with the display software. The tools do give the operator an intuitive interface.

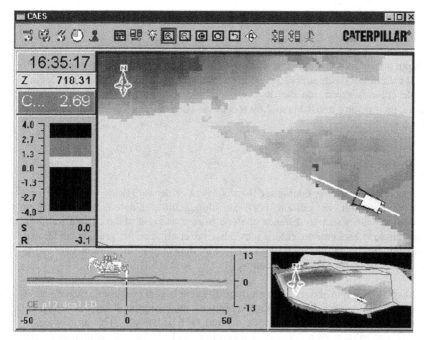

Fig. 4. CAES display screen. Multiple screens may be configured for each application

4.2   Plan View

CAES plan views display the current position and orientation of the machine on the design plan. As the machine moves and the terrain model is updated, the plan view is immediately updated with the new information. The plan view is color coded to show the operator exactly where the terrain needs to be cut or filled. Plan view colors may be changed, but in general once the display turns green, the terrain is within the predefined on-grade tolerance. The machine position is also continuously updated as GPS positions. Tools are available to zoom in or out, to rotate the display, or to center the machine in the display. Multiple plan views may be displayed in the same screen. This allows the operator to zoom in on a specific work area in one window while still displaying the entire plan.

4.3   Profile Window

The profile window also displays grade information. A cross-sectional view of the terrain either in the direction of heading or perpendicular to the heading direction is displayed. The profile window may be configured to display all active designs as well as the current elevation, and the original surface elevation. This means that information that may be hidden in the plan view, (which only shows the primary design) may be visible in the profile window. Profile traces are color-coded to so that each elevation layer is displayed in a different color. Tools are available to change the length of the profile and to toggle whether the profile is viewed parallel or perpendicular to the machine.

4.4   Depth Gauge

The depth gauge gives the operator a quick visual reference to determine the cut or fill value at the current machine location. This window is again color-coded. The default on-grade color is green and the on-grade tolerance is configurable. In the above example, the operator needs to cut 2.69 feet at the current point where the machine is working.

4.5   Digital Windows

Digital windows may be configured to display a wide variety of values. These include information on current position, slope, time, elapsed time, or even measurement values such as distance or slope from a set point to the current position. Digital window background colors may also be configured to provide additional information. For example, the background colors may display the current state of the GPS receiver or be set to cut or fill colors.

In the display example above, there are four different digital windows. One displays the current time, another the current elevation with the background color showing the current GPS accuracy (green indicates a fixed GPS solution). The third window shows the cut value with a cut color background, and the last one is displaying the current slope and roll values for the machine.

4.6   Productivity and Diagnostic Display Screens

Screens are also available to give operators productivity information. Data is available on volume of material moved on a design, cut and fill rates, and cycle times. Diagnostic screens may be used to provide basic troubleshooting support to operators or support personnel.

5   Machine Monitoring and Site Status in the Office

As noted in the data flow section, CAES continuously sends real-time terrain updates and machine positions to the office. This onboard information including the machine's topographical databases, position, productivity, and diagnostics, are transmitted from CAES on the machines to the office over the radio network. The terrain updates are incorporated by METScomms into the master site models. CAESoffice may then be used to view the current system state.

5.1   CAESoffice Functionality

CAESoffice is a virtual window into the construction site because it gives the user a real-time view to information collected onboard CAES-equipped machines.

CAESoffice is very flexible and has more functions than can be described in this document. Previous sections already outlined how CAESoffice enables construction planning personnel to preview elevation designs and compare them to the current topography before sending them to the machines. A few more examples will be provided in this section to give a glimpse of the many applications of CAESoffice.

Since CAESoffice can be distributed on a computer network, construction personnel in operations, planning, and management can all use CAESoffice to stay in touch with the current state of the construction. CAESoffice can deliver the information to the people who need it when they need it so that decisions can be made on a timely basis using current information.

5.2   Export of Site Status Information

In addition to viewing terrain data in CAESoffice, the current terrain may also be exported so that it may be used by site planning or development software. It is exported in a user defined ASCII file format to provide needed flexibility to the customer. Since CAES is constantly surveying the terrain, the exported data is a very complete and is an up-to-date model of the current ground. This data is then also the record of the as-built surface and may be used to document the completion of a project.

6   Machine Productivity

CAES on-board is responsible for deriving the productivity data from a number of different sources including GPS, the radio network, onboard sensors, onboard switches, and third party interface information which supply information from other on-board systems. The onboard system writes the data to a file, which is later sent to the office over the radio network. The onboard system is also responsible for displaying summary information to the operator's screen on the machine.

Each CAES application collects its own set of specific productivity data. Volume data is collected for all grade and slope applications. Machine utilization, which shows the amount of time the machine is working versus the time it is stationary or off, is generated for all machines. Track-type tractors collect cycle information. Other types of productivity information are available for other applications such as material control. The following sections provide examples of some cycle and volume reports.

Productivity information to the office may be scheduled through METSmanager so that updates occur on a regular interval. This productivity information will then be uploaded automatically to the productivity database where reports may be generated. All original files are archived for later reference.

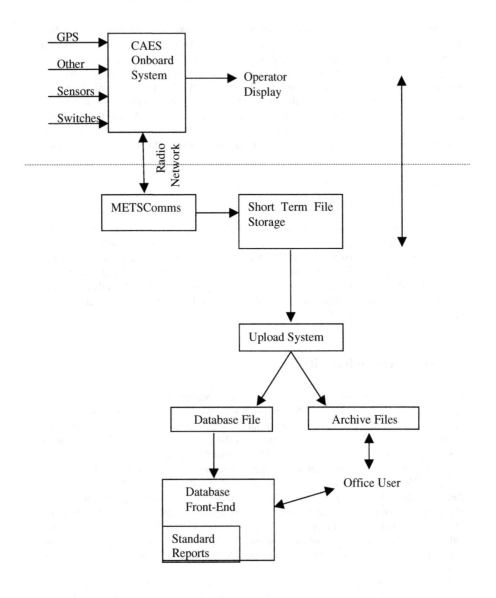

Fig. 5. Productivity Block Diagram

6.1   Productivity Information

Various kinds of productivity information are available on each machine type. Productivity information is available to both the machine operator through productivity screens and to office personnel through the files sent from the machines. These files are incorporated into the office productivity reporting tools.

6.2   Track-Type Tractor Cycle Information

A typical production cycle for a track type tractor consists of the machine digging, carrying the material, dumping the material, and then returning back to another dig point. Thus there is a forward leg of the cycle and a reverse leg. Calculations of cycle data include number of cycles since last reset; average cycles per hour; average cycle distance; average cycle slope and average forward and reverse speeds.

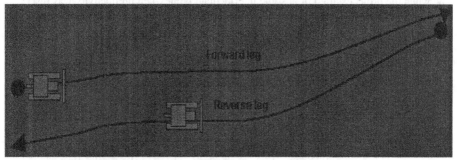

Fig. 6. Track-Type Tractor Cycle

6.3   TTT Cycle Productivity Report

This report is intended to compares the productivity of several machines using cycle data. The sample report quickly shows the average cycle times with forward time, reverse time, and stationary time for a fleet of machines. The table provides additional information on the average cycle times.

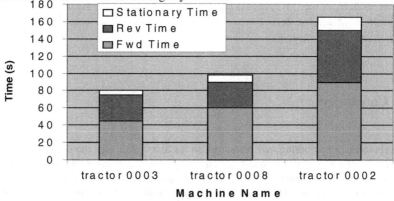

Mach Name	Total Cycles	AveDuration	Std Dev	MinDuration	MaxDuration
Tractor 3	50	80	3.3	70.2	90.7
Tractor 8	21	98	4.2	82.3	114.5
Tractor 2	14	155	14.7	112.5	189.9

Fig. 7. Sample Cycle Report

6.4   Volume Reports

Volume data is collected on the amount of material cut or filled. Volumes are calculated by comparing the current ground or original ground to the design. The original ground is simply a current ground archived at some point in time. This may be when the project started or at some intermediate point when productivity was reset. Volumes are calculated in real-time as the machine moves and the terrain is updated.

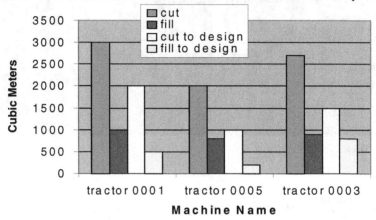

Machine Name	Average Fwd Vel	Average Rev Vel	Average Slope
Tractor 0001	3.1	4.3	.01
Tractor 0005	2	3.3	.025
Tractor 0003	2.7	4.0	-.01

Fig. 8. Sample Volume Report

7   Conclusion

The advances in computing and communications associated with the digital age have advanced earthmoving in construction to same magnitude that they have revolutionized other industries. With high technology information products such as Caterpillar's Computer Aided Earthmoving System, the earthmoving machines have become real-time, high accuracy, surveying instruments. Additionally, with wireless data transfer capability, these systems have streamlined the earthmoving process by providing the office personnel with real-time topographic updates and providing the transfer of design changes to machine operators in the field with no loss of productive time.

References

1. Greene, D., Caterpillar Inc., Concord, California: SME Preprint 97-164, Computer-Aided Earthmoving Systems-CAES. SME.
2. Harrod, G.R., Sahm, W.C., Caterpillar Inc., Peoria, Illinois: SME Preprint 98-113, Mining Information Goes "Real-Time". SME.

Feature Lines Reconstruction for Reverse Engineering

Alessandro Raviola, Michela Spagnuolo, and Giuseppe Patané

Istituto per la Matematica Applicata,
Consiglio Nazionale delle Ricerche,
Via De Marini 6, 16149 Genova, Italia
{spagnuolo,patane}@ima.ge.cnr.it
http://www.ima.ge.cnr.it

Abstract. This paper addresses the reconstruction of an object shape model from a set of digitized profiles, or scanlines. The reconstruction is approached in two main phases. Firstly, a hierarchical simplification of the original data set is performed which is aimed at discarding irrelevant data and at providing different levels of detail of the data set. Secondly, a shape signature is computed to characterize the shape of each profile and to reconstruct important feature lines. Feature lines can be used to delimitate meaningful surface patches on the reconstructed mesh (segmentation). Even if the proposed approach is presented in the specific context of Reverse Engineering, its application and usefulness is more general as it will be discussed for the geographical domain.

1 Introduction

Issues related to surface reconstruction occur in a variety of applications: digital terrain modelling, surface reconstruction from range or laser scanner data, interactive surface sketching. The methods developed have been mainly defined on a case by case manner to exploit partial structures on the data: for instance, algorithms for reconstructing surfaces from contours make heavy use of the particular spatial distribution of the data [22]. Methods developed for unorganized points, that is, methods which do not make any assumption on the spatial distribution of the data, generally use neighbouring properties of the points for reconstructing locally the surface in a piece-wise fashion [1,6,9]. The type of surface to be reconstructed usually depends on the application context: for example, in digital terrain modelling, a mesh-based interpolation of the data set is generally considered satisfactory while in reverse engineering more complex surface representations may be required.

Reverse engineering is an interesting approach to object design and production: dense samplings of real parts or prototypes are transformed into CAD models, thus providing a great flexibility to the design phase. Reverse engineering enables to produce objects when original drawings or documentations are unavailable and it makes also possible to re-engineer an existing part after analysis and/or modifications [12,23]. At a very general level, the reverse engineering

C.Y. Westort (Ed.): DEM 2001, LNCS 2181, pp. 18–30, 2001.
© Springer-Verlag Berlin Heidelberg 2001

flow starts with the acquisition of sampled points on the object surface, using non-contact methods, such as laser or magnetic fields, or tactile methods, such as mechanical probes touching the surface along predefined paths. The acquired data are transformed into an intermediate geometric model, usually a triangular mesh, which is further processed in order to obtain a CAD model of the object using different methods to achieve freeform surface patches. Therefore, the main purpose of reverse engineering is to convert discrete data sets into piecewise smooth, continuous models. A basic step towards the creation of a CAD-like model is the segmentation which provides a grouping of the data or a decomposition of the intermediate model into subsets belonging to elementary surface types [12,13,23].

The approaches used for the segmentation process are general or dedicated. The first ones use only a general knowledge of the surface to execute the segmentation while dedicated approaches, which are preferable, search for particular structures related to the application's environment. For example, it might be preferable to represent a cylindrical hole by one functional face to be manufactured by grinding or drilling, instead of approximating it by facets. Therefore, the decomposition into functional patches for CAD/CAM applications is one of the main topics in this research area [23]. Furthermore, surface decomposition is used in spatial data handling for the automatic extraction of morphological features and for the classification and characterization of features of the terrain relief. Therefore, many common issues and problems may be found in reverse engineering and digital surface modelling: the reconstruction of a geometric model, which interpolates the samples, and the recognition of meaningful subsets in the geometric model.

While we firstly approached these problems in the context of digital terrain modelling [20], the work here presented is mainly concerned with an application to reverse engineering. The aim is to define a high-level polyhedral model of the object surface, as a triangulation constrained to feature lines [15,19]. Assuming that data are distributed along vertical sections, corresponding to scanlines, our approach is described by the following steps:

- *data simplification with a multiresolution technique*: this aims at reducing the samples to a minimal number discarding those which do not locate important shape features;
- *profile analysis*: geometric analysis of each simplified profile to classify its points according to their contribute to define the profile shape;
- *feature lines detection*: extraction of feature lines, or key curves, achieved by joining points which lie on adjacent scanning lines and which are judged to be similar according to the measures defined in the previous step.

The purpose of simplification is the reduction of sampled points to prevent the construction of bulky models whose large amount of data is usually a bottle neck for the analysis phase. The simplification is performed line by line, exploiting the spatial distribution of samples along planes, and the method used ensures to keep the simplification error within a user-defined range. This step is obviously relevant to the geographic domain, where the amount of data needed to faithfully represent a terrain is very big, and the proposed approach may be used for simplifying profiles as well as contour lines.

Once the scanning lines are reduced to the required levels of detail, a multiresolution representation of the lines is used to distinguish between points locating global or local features on the profiles. The scale and two geometric measures characterize the shape feature induced by each point on the simplified scanline. In this application the geometric measures are finalized to the classification and extraction of features meaningful in the CAD/CAM environment, such as slots or pockets, but the classification might be changed in order to fit other application domains. Finally, the object feature lines are obtained linking points that lye on adjacent scanning lines and which are judged similar according to the defined measures.

A first version of this process was applied to the description of seafloor relief, where bathymetric profiles, seen as vertical sections of the surface, were analyzed to extract and characterize elongated features such as ridges and ravines [18]. Using the scale-based description of each profile, a schematized representation was defined to characterize the shape of the seafloor around characteristic lines.

The paper is organized as follows. First, the simplification technique, which is a 3D extension of the tolerance band method, is presented in section 2. Then, the measures used to classify the shape features of the scanning lines are introduced in section 3. Finally the feature lines reconstruction and considerations about the segmentation process are given in section 4.

2   Simplification

New acquisition technologies have accelerated, on the one hand, the manufacturing process providing dense data set in order to achieve accurate reconstructed products and have brought, on the other hand, the need of discarding redundant information. Therefore, it is needed to reduce the input data set maintaining points with a high information content, preserving the shape and controlling the approximation error [8,16]. The full automation of the simplification process, which has been addressed by several authors [3,4,17], represents an open issue.

First of all, the sampled object is assumed to be a 2.5D surface with points distributed along slicing planes, which define a sequence of scanning lines, whose definition is given below.

Definition. Let be $F := \{(x, y, z) \in \mathbb{R}^3 : f(x, y) = z, (x, y) \in D\}$ the graph of a single-valued function $f : D \subseteq \mathbb{R}^2 \mapsto \mathbb{R}$ which defines the object surface and $S := \{(x, y, z) \in \mathbb{R}^3 : ax + by + c = 0\}$ a plane ($a \neq 0$ or $b \neq 0$). A scanning line L is an ordered sequence of $(n + 1)$ points $L := \{P_0(x_0, y_0, f(x_0, y_0)), \ldots, P_n(x_n, y_n, f(x_n, y_n))\}$ such that $L \subseteq F \cap S$ and at least one between $(x_0, \ldots, x_n)$ and $(y_0, \ldots, y_n)$ is a closely increasing or closely decreasing sequence. Then, the sampling of F with respect to the scanning direction defined by the normal vector $(a, b, 1)$ to a set of planes $S_i = \{(x, y, z) \in \mathbb{R}^3 : ax + by + c_i = 0\}$ $i = 0, \ldots, k$ is represented by $L_0, \ldots, L_k$ where each L_i is the scanning line on F with respect to S_i.

At this stage, we will assume that measurement errors introduced during the acquisition have been already analyzed. This means that only spikes and noise points, which can be identified in a simple way, have been processed and

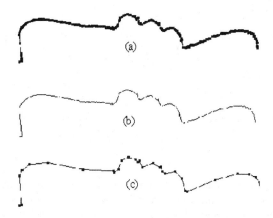

Fig. 1. Original scanning line (376 points) rendered with points (a), without (b), and the simplified line (31 points). The reduction rate is about 92%.

eliminated. Global filters (i.e. median, averaging, gaussian filters), if required, can be applied at the end of the feature lines extraction avoiding problems in the identification of features. See section 4 for more details.

The aim of line simplification is the reduction of the points number of L, that is, the definition of a new line L' with fewer points such that $L' \subseteq L$. This is also due to our application constraint, that is, the requirement of avoiding points displacement, which forces us to select points in the original set only and prevents us to use wavelet-based techniques which do not guarantee this property [20].

Therefore, the simplification method is based on a 3D extension of the Douglas and Peucker algorithm [5,10,15,19,21] commonly used for cartographic applications. The basic idea of this method, which produces the least area and vector displacement from the original line [14], is the use of a tolerance band whose width is related to the scale of detail which can be considered irrelevant. Given a tolerance band α_0 (see fig.4), the polyline at issue is approximated by:

- the line segment $\overline{FL}$ if the farthest point P^* from this segment is less or equal to α_0, otherwise
- we will split the chain in P^* applying the criterion recursively to the new polylines FP^* and P^*L.

Selected points are finally chained to produce the simplified scanning line. In fig.1, it is shown an example of the application of this algorithm with a given tolerance band. The original scanning line (fig.1(a) and 1(b)) is composed of 376 points and the simplified one (fig.1(c)) has only 31 points, with a reduction of 92%.

In our application, the algorithm has been implemented using a tolerance cylinder instead of a rectangle so that 3D curves can be simplified [19,20]. Our

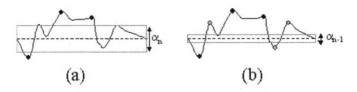

Fig. 2. The points remaining with an α_n band (a), and those (b) remaining with the use of α_{n-1}.

approach to line simplification is twofold: it aims at reducing the number of points and also at evaluating the scale of the shape feature introduced by each point. In this sense, this algorithm enables a classification among points of local and global importance using a multiresolution approach [1]. This model is based on the use of a set of tolerance bands whose size varies in the α-set $\{\alpha_0, \ldots, \alpha_n : 0 < \alpha_0 \leq \ldots \leq \alpha_n, n \geq 1\}$. Each element of the α-set is called α-width and it represents the value of a tolerance band. The minimum width α_0 is chosen in relation with the maximum acceptable error for the simplification, while α_n corresponds to the dimension of the maximum enclosing rectangle containing the original line. Given a line in the input data set, L' will indicate the simplified scanning line with the width α_0.

Applying the Douglas-Peucker algorithm to L' using the greatest band α_n the set P_n will be defined (see fig. 2(a)). Using the α_{n-1}-width we will find a new set P_{n-1} such that $P_n \subseteq P_{n-1}$, since $\alpha_{n-1} < \alpha_n$, (see fig.2(b)). Therefore, applying the whole α-set to L' we obtain the sequence $P_i \subseteq P_{i-1} \; \forall i = 2, \ldots, n$ where P_i is the set of points detected with the tolerance band α_i.

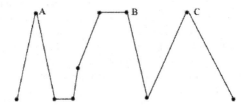

Fig. 3. A simplified scanning line with different shape features.

Our simplification method starts using the largest α_n-width of the α-set and labelling the points in P_n with the width α_n. The algorithm is applied again with band α_{n-1} and the remaining points, not previously labelled, are marked

---

[1] It is possible to set up a model for the multiresolution model and to evaluate the error induced in the simplification process with the use of the Hausdorff distance. For further readings we refer to [15,21].

with this new value. The algorithm is iteratively applied with decreasing values in the α-set, and, at the end, the classification of the points will reflect the contribute of each point to the overall line shape. Therefore, points found with a greater tolerance band will locate global features while points found with a smaller value are related to local shape elements. The accuracy of the process is strictly related to the width of the smallest band α_0 and all lines have to be simplified using the same α-set of tolerance bands. Problems may arise due to the unbalanced density and dimension of the shape features. The α-set can be set up considering the largest bounding box α_n, the minimum levels of detail α_0 and defining the other tolerance bands using an uniform distribution

$$\alpha_i := \alpha_0 + \frac{\alpha_n - \alpha_0}{n} \times i \qquad i = 1, \ldots, n.$$

This choice may create problems when the shape features of a line are very different: a possible solution can be achieved using a re-calibration of the α-set.

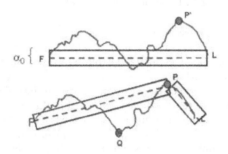

Fig. 4. The points remaining with an α_n band (a) and those remaining with the use of α_{n-1}.

3   Scanning Line Analysis

The aim of scanning line analysis is the classification of shape features and it represents the basis for the reconstruction of features. This process relies on the assumption that the object is well sampled, that is, we assume that the shape of adjacent scanning lines is similar. Feature lines of the object will be constructed as a set of similar points: therefore, it is necessary to classify points of the simplified lines in order to be able to evaluate shape similarity among them. As described in the previous section, using the Douglas-Peucker method with a user defined α-set it is possible to obtain a first classification of points with respect to the size of the shape features they introduce [15,19].

Because of the scale is not sufficient to classify a linear shape other geometric parameters have to be considered. For example, in fig.3 line is shown where

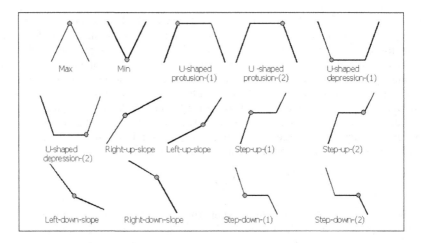

Fig. 5. Basic shape elements classification.

points are classified at the same scale but whose shape is different; points A and C are similar but point B differs. Therefore, we can say that A and B have the same information with respect to shape size even thought they are related to shape features of different type. Two measures are introduced to distinguish different shape features; the first one used to code the type of feature and the second one, based on angular parameters, to locate its orientation in space. These measures characterize the curvature and the local orientation of linear shapes [2,7,11]. Since the number of points which describes a feature is three or four, we have adopted a classification scheme which uses at most three points in the neighborhood of the focus point (all the possible configurations are shown in fig.5 and described in table 1). Moreover, with this classification we have tried to identify shapes which are meaningful for producing a decomposition which is relevant with respect to basic features in CAD applications. This choice provides flexibility especially for the recognition of features and it will be discussed in section 4. For each configuration, the square represents the point at which the label will be linked. Table 1 can be improved using different geometric criterions for the identification of other shape elements introducing a possible application of the model for geographical information systems and for a more general pattern recognition process.

The primitive shape elements described in fig.5 are defined in table 1. We underline that the object surface is assumed to be a height field (2.5D), and that $A, B, M, C \in \mathbb{R}^3$ are points used to identify a given configuration, where M is the focus point. Finally, $det(A, M, B)$ is the determinant of the matrix used to indicate if AMB is a left or a right turn on the slicing plane.

The second classification is based on angular measures and it is necessary to distinguish among shape features (fig. 7,9). Three angular measures are introduced considering the same triplet of points as before. In fig.6, the geometric

Table 1. Mathematical description of the primitive shape elements. The equality relations have to be considered "at most equal" with respect to the scanning error.

Description	Properties
Max (V slot reversed)	$z_A < z_M, z_M > z_B$
Min (V slot up)	$z_A > z_M, z_M < z_B$
U shaped protusion 1 (U slot reversed 1)	$z_A < z_M, z_M = z_B, z_B > z_C$
U shaped protusion 2 (U slot reversed 2)	$z_A < z_B, z_M = z_B, z_M > z_C$
Step up 1	$z_A < z_M, z_M = z_B, z_B < z_C$
Step up 2	$z_A < z_B, z_B = z_M, z_M < z_C$
U shaped depression 1 (U slot 1)	$z_A > z_M, z_M = z_B, z_B < z_C$
U shaped depression 2 (U slot 2)	$z_A > z_B, z_M = z_B, z_M < z_C$
Step down 1	$z_A > z_M, z_M = z_B, z_B > z_C$
Step down 2	$z_A > z_B, z_M = z_B, z_M > z_C$
Left up slope	$z_A < z_M, z_M < z_B, det(A, M, B) > 0$
Right up slope	$z_A < z_M, z_M < z_B, det(A, M, B) < 0$
Right down slope	$z_A > z_M, z_M > z_B, det(A, M, B) < 0$
Left down slope	$z_A > z_M, z_M > z_B, det(A, M, B) > 0$

representation of the values used to quantify this measures is given for a right-up-slope feature and according to table 1.

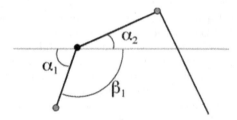

Fig. 6. Configuration type right-up slope and its characteristic angles.

The angles evaluate the amplitude and the orientation of the feature. For example, the amplitude at point M_1 is given by the angle β_1 defined by the line segments AM_1 and M_1B (see fig.7, 9). Similarly, the orientation of the shape feature at M_1 is given by the angles α_1 and α_2. For each shape feature, these two latter angles are computed by considering a horizontal line of reference and are evaluated as the minimum angles between the reference line and the segment defined by the focus point and the previous (resp. next) one. Using the α-set, the previous classification and the angles previously defined we are able to introduce three basic similarity "relations" on the input data set:

– α-width similarity: $P \in L_P, Q \in L_Q$

$$P \sim_\alpha Q \Longleftrightarrow \text{width}(P) = \text{width}(Q)$$

where width(K) is the α-width that characterizes the point K in its scanning line,
- feature similarity: $P \in L_P, Q \in L_Q$

$$P \sim_T Q \Longleftrightarrow T(P) = T(Q)$$

where $T(K)$ represents the class of point K according to table 1,
- geometry similarity: $P \in L_P, Q \in L_Q$

$$P \sim_G Q \Longleftrightarrow \text{their orientation and inner angles are similar}$$

(i.e they differ less than a given threshold).

4   Feature Line Detection

We underline that segmentation can be developed in different ways which depend on the spatial organization of sampled data introduced by scanning techniques. For instance, our approach try to extract feature lines starting from a set of scanning lines exploiting the neighboring information implicitly defined by points connectivity. Its application is therefore restricted to a specific structure in the input data set. Other methods [6,9] produce an arbitrary surface segmentation where the patches decomposition are based on the simplification of an underlining triangulation. This approach is not suitable for reverse engineering applications because important shape elements, such as sharp edges and lines, can not be identified. Other methods use a discrete curvature approximation to segment the surface into mechanical meaningful patches [23], introduced by scanning techniques. These methods are more general, with respect to the organization of the input data set, than our model but they require a strong preprocessing (i.e simplification, smoothing, etc., etc.) of the given polyhedral surface where sharp elements may be deleted or changed loosing important shape information for the segmentation process.

Beside shape characterization, the measures suggested (i.e. α-set and angles) for the evaluation of the similarity are intended for the reconstruction of feature lines using the similarity relations previously defined. Indeed, we are trying to recognize automatically feature lines in a direction almost orthogonal to the scanning direction, by linking similar points from one scanning line to the next one.

More precisely, the detection of the feature lines can be explained as follows: starting from a feature point, similar points are searched in the adjacent scanlines which are selected as candidates for joining. The distance among points is considered because the reliability of the line reconstruction decreases when the first one increases. This is achieved using an influence cone [20] with a variable axis oriented in the direction of the line which we want to reconstruct, is used to select parts of adjacent lines that are relevant for the reconstruction. Among points falling within the influence cone, one or more similar points are selected and linked to the point at issue. The similarity of two points is based on the

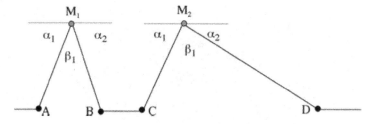

Fig. 7. Angles used to classify shape features.

relations previously introduced: two points P and Q can be joined with a line segment if

$$P \sim_\alpha Q \text{ and, } P \sim_T Q, \text{ and } P \sim_G Q.$$

The three relations can be used with the same weight, meaning that each measures equally contribute to the identification of feature, or different weights can be associated. This means that, using a set of weight W and some knowledge about the surface, it is possible to improve the quality of the feature lines detection. For example, we can decide a different set of weights for a natural surface, like a terrain or a seafloor, with respect to a synthetic surface like that of a manufacturing object taking into account its morphology and regularity. Moreover, different weights can be applied to an iterative analysis of the same surface, for example giving more importance to the scale in the first step and to geometry in the second one. Obviously, the scanning direction is fundamental for the feature lines detection process, as those ones whose direction is almost parallel to the scan direction will be, in general, impossible to detect with this approach.

In general, a point can be connected to more than one point giving rise to a split of the feature lines (see fig.8). Another difficulty faced during the reconstruction is the problem of feature lines intersection which should not be in principle admitted beside specific situations.

5   Future Work and Conclusion

The main contributions of this paper concerns:

- the organization of the simplified data set in a multiresolution scheme which allows us to distinguish among global and local features of each profile,
- the use of similarity measures to automatically extract feature lines from the set of scanning lines.

Since each scanline is fully characterized by the scale and shape of its points, it turns out that scanlines can be decomposed in parts corresponding to meaningful shape features of the object model. This is one step towards the automatic segmentation of the reduced data set into patches delimited by the extracted feature lines. Moreover, we believe that the multiresolution representation of the scanning lines will allow us to detect simple and compound features

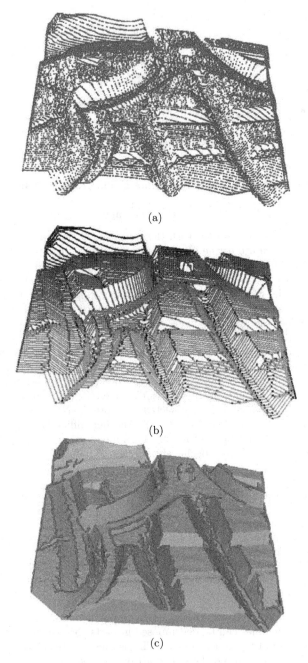

(a)

(b)

(c)

Fig. 8. Example of segmentation applied to a boot sole: (a) input object, (b) points classification (c) feature lines reconstruction.

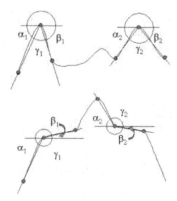

Fig. 9. Classification of features based on angular measures.

(e.g. nested slots), therefore providing a manufacture-oriented segmentation of the data set. Future work will mainly consist in the full development of the segmentation step. With respect to previous approaches, based for example on wavelets [20], the hierarchical use of tolerance bands does not cause any smoothing effect or points displacement avoiding the smooth of sharp edges in the samples. Moreover, another advantage we foresee is the correction of samples using the extracted feature lines, mainly for straightening edges or correcting almost perpendicular faces, which may be slightly distorted by the sampling step.

Acknowledgements. The authors would like to thank the Technimold S.r.l., Genoa-Italy for the fruitful co-operation during the Project "Definition of New Technologies for Reverse Engineering" and for the data provided for this work. Special thanks are given to Dr. Corrado Pizzi, IMA-CNR, for the valuable help and support.

References

1. Attene, M. Spagnuolo, M.: Automatic surface reconstruction from point sets in space, Computer Graphics Forum, (2000), 19(3)
2. Atteneave, F.: Some information aspects of visual perception. Psycological Review, Vol. 61, n. 3 (1954), 183-193.
3. Ballard, D.H.: Strip-trees: A hierarchical representation for curves. Communication of the Association for Computing Machinery, 3 (1986), 2-14.
4. Buttenfield, B.P.: A rule for describing line feature geometry. In B.P. Buttenfield R. McMaster (eds) Map generalization, Chapter 3, (1997), 150-171.
5. Douglas, D.M., Peucker, T.K.: Algorithms for the reduction of the number of points required to represent a digitized line or its caricature. The Canadian Cartographer, 10/2, 1973, 112-122.
6. Eck, M., Hoppe, H.: Automatic reconstruction of B-spline surfaces of arbitrary topological type. Computer Graphics, SIGGRAPH' 96, August 96, New Orleans, Louisiana, 99-108.

7. Falcidieno, B., Spagnuolo, M.: Shape abstraction tools for modeling complex objects. Proc. of the International Conference on Shape Modeling and Applications, 1997, IEEE Computer Society Press, Los Alamos, CA.

8. Fisher, A.: A multi-level for reverse engineering and rapid prototyping in remote CAD systems. Computer Aided Design 32 (2000), 27-38.

9. Guo, B.: Surface reconstruction from point to splines. Computer-Aided Design, Vol. 29, n. 4, (1997), 269-277.

10. Hershberger, V., Snoeyink, J.: Speeding up the Douglas Peucker line simplification algorithm. Proc. 5th Intl. Symp. on Spatial Data Handling, Vol.1, Charleston, SC, (1992), 134-143.

11. Hoffman, D., Richards, W.: Representing smooth plane curves for visual recognition. Proc. of American Association for Artificial Intelligence, Cambridge, MA: MIT Press (1982), 5-8.

12. Hoschek, J., Dietz, U., Wilke, W.: A geometric concept of reverse engineering of shape: approximation and feature lines. In: M. Daehlen, Lynche T. and Schumaker LL. (eds): Mathematical methods for curves and surfaces II, Vanderbilt University Press,(1998) Nashville.

13. Jiang, X.Y., Bunke, H.: Fast Segmentation of Range Images into Planar Regions by ScanLine Grouping. Machine Vision and Applications, Vol. 7, No. 2, 1994.

14. R. McMaster. A statistical analysis of mathematical measures for linear simplification. The American Cartographer 13, 2 (1986), 103-117.

15. Patané, G., Pizzi, C., Spagnuolo, M.: Multiresolution compression ad feature lines reconstruction for Reverse Engineering. Proc. of the 5th Central European School on Computer Graphic, CESCG2001, April 2001, Bratislava 2001, 151-162. URI: http://www.isternet.sk/sccg/main_frames.html.

16. Patrikalakis, N. M., Fortier, P.J., Ioannidis, Y., Nikdaon, C.N , Robinson, A. R., Rossignac, J. R., Vinacua, A., Abrams, S. L.: Distributed Information and Computation in Scientific and Engineering Enviroments. M.I.T 1998.

17. Plazanet, C.: Modelling Geometry for Linear Feature Generalization. In: M. Craglia, H. Coucleis (eds), Bridging the Atlantic, Taylor & Francis (1997), 264-279.

18. Plazanet, C., Spagnuolo, M.: Seafloor Valley Shape Modelling. Proc. of Spatial Data Handling, Vancouver, (1998).

19. Raviola, A., Spagnuolo, M.: Shape-based Surface Reconstruction from Profiles for Rapid/Virtual Prototyping. Proc. of Numerisation 3D, Paris 1999.

20. Spagnuolo, M.: Shape-based Reconstruction of Natural Surfaces: an Application to the Antarctic Sea-Floor. In M. Craglia, H. Coucleis (eds), Bridging the Antarctic, Taylor & Francis (1997).

21. Saux, E.: Lissage de courbes pur des B-splines, application á la compression et á la généralisation cartographique. PhD Thesis, Nantes University, France, January 1999.

22. Tang L.: Automatic Extraction of specific geomorphologic elements from contours. Proc. of Spatial Data Handling, Charleston, SC, USA (1992) 554-566

23. Varaday, T., Martin, R., Cox, J.: Reverse engineering of geometric models, an introduction. Computer Aided Design, Vol. 29, No. 4, (1997), 255-268.

Geomorphometrical Mapping of Relief-Dissection Using GIS

Gábor Mezősi and Richárd Kiss

University of Szeged, Department of Physical Geography, POB 653,
H-6722 Szeged, Hungary
Tel./Fax.:36 62 454158
{mezosi, ricsi}@earth.geo.u-szeged.hu

Abstract. In this project we have made an attempt to find out the limits of GIS in morphometrical analysis. The aim is not only to produce and analyse a vertical relief-dissection map, but we wish to point out the regional differences in erosion and its translocation in a geological scale. According to our idea, the vertical relief-dissection map (i.e. negative relict surface) can be constructed for each stream order by subtracting the real surface from the summit planes fitting on the watersheds. Such maps can give information about long-term translocations of erosion and about the changes of its rate. The mapping of vertical relief-dissection is based on morphometrical analyses of the 70's, but as there where no suitable methods it was carried out only under limited conditions. The surface modelling of GIS has opened a new way in this direction.

Introduction

Studying the erosion development of a given surface is a traditional subject in geomorphology. In the latest 20 years geomorphology has moved away from the investigation of the denudation chronology towards the study of processes. The traditional problem of geomorphology is the connection between shapes and processes. The explanation of land surface and the calculation of the characteristics of surface need quantitative description of the relief. In the 70's, morphometry gave a solution for the problem by quantification of the relief. From the middle of the 80's, GIS has proved to be a very powerful and useful method and offered advantages for spatial distribution of geomorphological processes.

The landscape by its work often leaves measurable marks behind. From the point of view of the destroying effect, of the intensity of erosion, it is a good estimate to measure the amount of material that has been transported away. Starting from the flat surface the rate of relief transformation can be deduced from the dissected river network, or from the amount of material that has been carved out from the flat terrain and taken away by the rivers. This volume can be calculated by morphometric analyses step by step according to the dissection rate of the river network's. The degradation is negligible on the edges of the catchment area. If we set a plane on these catchment-edges, we can call this surface the "summit plane" compared to the present situation.

C.Y. Westort (Ed.): DEM 2001, LNCS 2181, pp. 31–38, 2001.
© Springer-Verlag Berlin Heidelberg 2001

Method

We have analysed an approximately 100 km^2 catchment area. It is situated on the NE edge of the 15 million-year-old Mátra volcano (in Northern-Hungary), which produced mostly andesite and riolithe. Therefore, since our test area was more or less geologically homogeneous, we cannot take into account the possibility of translocation of stream directions. In addition, the precipitation conditions are very similar all over the area, thus we do not suppose significant change in the rate of erosion.

We have supposed that a currently six-ordered stream (Strahler-order) was five-ordered earlier, four-ordered even before, and so on. This works only as a model, because in spite of the homogeneity geological, climatological and orographical differences could occur. The usability of the order system for statistical/GIS investigation was also supposed. The aim of the research is to highlight these disturbances which can be investigated in the drainage-network.

First of all, we have classified the stream network of the area according to Strahler's system (Fig. 1). The stream network was defined by using the Multiple Flow Direction (MFD) (Freeman 1991) and the Deterministic 8 (D8) (Tarboton et al. 1991) models. Both models calculate the orographically definable catchment area for each pixel in the area. While the D8 model allows water to flow only toward one of the neighbouring pixels, MFD tolerates multiple flow directions as well, i. e. toward more than one pixel.

When calculating the entire catchment area, we have used the real surface, since this is larger than that of the area shown in the map's projection. The real pixel area was calculated with the following formula:

$$T_f = T_c / \cos(S),$$

where T_f stands for actual (absolute) surface area, T_c for the area of a pixel and S indicates slope.

In case of low accumulation values, we have used the MFD model, providing a correct result with non-convergent territorial flows. Where the accumulation value exceeded a threshold of 25000 m^2 (set by experience) we used D8, which is more suitable for modelling linear erosion processes. By extracting high values from this map (over 75000 m^2 in this case) we can produce the stream network, based on which we can define the individual partial catchment areas. We have counted more than 1000 stream segments. The large number of streams confirmed the statistical results of the drainage analysis.

Secondly, we have subtracted the real surface heights from the summit planes fitting on the watersheds of first-, second-, etc. order. In this way we got different vertical relief dissection-map for each order using ArcInfo 7.0.3. GIS. In order to prepare the n-order vertical variance map, we have created a surface using the intersections of the contour lines and the n-order catchment area boundaries (height values are derived from the values of the intersecting contour lines). The original method considered this as the summit plane (Kertész 1974). Based on GIS analyses, we have concluded that this method is only valid for ideal catchment basins. In most cases the development of the catchment area is far from ideal which makes the real surface to come above the calculated summit plane, thus negative values can appear

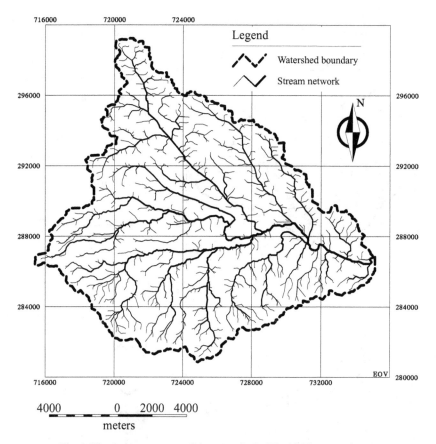

Fig. 1. The drainage system of the watershed of Parádi-Tarna creek

on the vertical dissection map. This is opposite to our goals. To solve this an iteration method has to be applied on the data as described below:

1. We create the summit plane according to the above method with Deluneay-triangulation taking the triangle corners from the watersheds.

2. If the minimum difference between the summit plane and the original surface (vertical variance) is lower than a custom threshold value (depending mostly the precision of elevation data) the result is ready, otherwise the third step has to be done as well.

3. Using the place and elevation value of the minimum point (as well as all other points) we create the new summit plane and submit it to the second step again.

It is advisable to examine the elevation values which were used during the iteration procedure ("positive anomalies"). It is most likely that these points and their surroundings differ from their environment concerning geological and geomorphological characteristics.

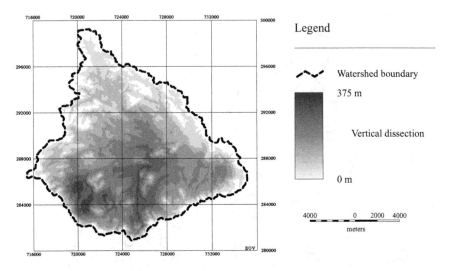

Fig. 2. 6th order relief-dissection of the test area

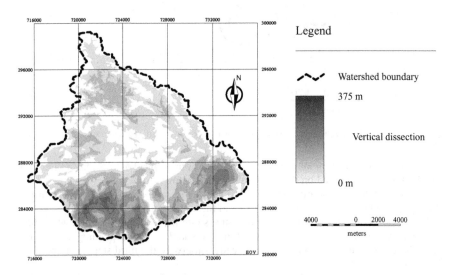

Fig. 3. 5th order relief-dissection of the test area

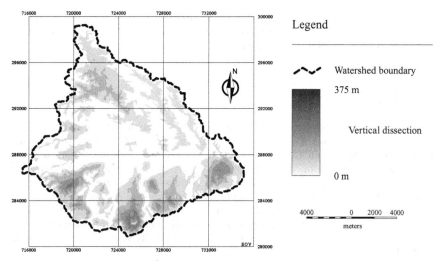

Fig. 4. 4th order relief-dissection of the test area

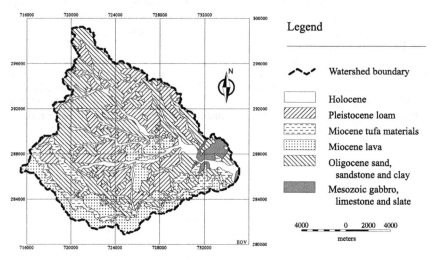

Fig. 5. The Geological map of the test area

When comparing the resulting surfaces, it has to be noted that the error possibilities of both calculated surfaces can add up thus decreasing precision. Care has to be taken when comparing the difference between the first order summit plane and the present surface to other differences, because while the different order summit planes are derived from exponentially linked valley networks, the relation of the present surface and the first order summit plane depends on the criteria defining our valley choice.

As the river network develops and grows with time it is possible to define lower and lower order catchment areas (e. g. those of tributaries) in a way similar to the above. Theoretically, this can be performed for each and every creek. By analysing the spatial differences of the intensity of this process we can obtain new kind of information about relief transformation.

Results

The vertical relief dissection-maps belonging to the different orders can be seen on Fig. 2-4. Their analyses show that two "proto"-rivers started to dissect the surface from south to north and from north-west to south-east, following the main slope directions. One was formed on the southern part of the catchment on Miocene andesite and riolithe lava, the second one was cut into a loose Oligo-Miocene sandy material (Fig. 5). The sinking and tilting of the Mátra resulted regional differences in the rate of erosion: on the southern part the small basins were formed bordered by dykes, and the relative relief has not increased significantly; in the northern part the relative relief grew considerably, so the erosional intensity became greater than on south (Fig. 4).

Another important change in fluvial erosion could happen during the Pliocene, when central part of the Mátra was uplifted by about 150-200 m. Therefore, the erosional-rate increased in the southern part of the catchment being closer to the centre than to the northern part.

It is well-known that there is an exponential relationship between the stream orders and the lengths of certain stream orders. The Fig. 6 shows that the number of the five-ordered streams (as well the six-ordered ones) is smaller than the expected. Their mean lengths and the volumes (Fig. 7) belonging to the vertical relief-dissections are decreasing radically compared to the statistical exceptions. It can be explained partly by lithological reasons: the five-ordered rivers situated   in the middle of the catchment, in an erosion resistant environment, and partly by the decreasing relative relief.

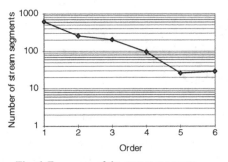

Fig. 6. Frequency of the stream segments

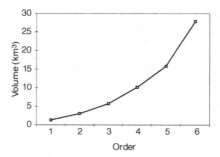

Fig. 7. Volume of the relief-dissection

The above mentioned geological and orographical influences can be studied along one valley. The Fig. 8a shows the vertical relief-dissections of different ordered reaches of that valley, which running from south to north and then turning to east. The disturbance of the curve of the vertical relief-dissection, for example which belongs to the five-ordered reach - at 13,350 m far from its beginning-, can be explained by geological reason: here the valleys reached the lower lying lava layers, therefore the drainage pattern became sparser. The regular rhythm of the curves can be plotted as semi-circles (Fig. 8b). The integration of the changes in erosional-rates into the exact geological time-scale is not yet solved.

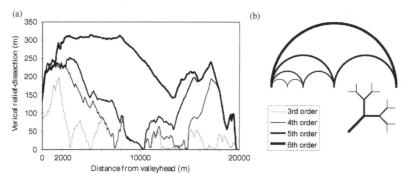

Fig. 8. (a) Relief-dissection along the Ilona-valley. (b) Rhythm of the relief-dissection in an ideal network.

References

Church, M.: Channel morphology and typology. In: The rivers handbook. eds. Callow, P. Blackwell, Oxford (1992) 126-143

Costa-Cabral, M. C., Burges, S. J.: Digital elevation model networks (DEMON): A model of flow over hillslopes for computation of contributing and dispersal areas. Water Resources Research Vol. 30, no. 6 (1994) 1681-1692

Evans, I.: What do terrain statistics really mean? In: Landform monitoring, modelling and analysis. eds Lane, S., Richards, K. John Wiley, Chichester (1998) 119-135

Freeman, T. G.: Calculating catchment area with divergent flow based on a regular grid. Computers and Geosciences Vol. 17 no. 3 (1991) 413-422.

Kertész, Á.: A morfometria és a morfometrikus térképezés célja és módszerei. In: Földrajzi Értesítő Vol. 4. (1974) 433-442

Montgomery, D. - Dietrich, W. The role of GIS in watershed analysis. In: Lane, S. - Richards, K. Landform monitoring, modelling and analysis. John Wiley, Chichester 241-262

Moore, I. D., Grayson, R. B. and Ladson, A. R.: Digital Terrain Modelling: A Review of Hydrological, Geomorphological, and Biological Applications. Hydrological Processes Vol. 5 (1991) 3-30

Strahler, A. N.: Quantitative geomorphology. In: Fairbridge, A. Encyclopedy of Geomorphology, New York (1968)

Tarboton, D. G., Bras, R. L., Rodriguez-Iturbe, I.: On the extraction of channel networks from digital elevation data. Hydrological Processes Vol. 5 (1991) 81-100

Zevenbergen, L. W., Thorne, C. R.: Quantitative Analysis of Land Surface Topography. Earth Surface Processes and Landforms Vol. 12 (1987) 47-56

AutoCAD[1] Land Development Desktop Release 2i

Michael Schoenstein

Autodesk GmbH, Hansastr. 28,
80686 Munich, Germany
michael.schoenstein@autodesk.com

Abstract. Land development projects require the skills of numerous design professionals: land and urban planners, surveyors, civil engineers. Whatever your contribution, whatever the size of your project, you need **a CAD solution** with topographic analysis functionality, real-world coordinate systems, volume totals, roadway geometry—all in an environment that promotes collaborative design and analysis. And you need to customize your CAD solution to fit the way you work. You can have it all with **AutoCAD® Land Development Desktop Release 2i** software—your platform for success.

The AutoCAD for Land Development

AutoCAD Land Development Desktop 2i extends AutoCAD Map® 2000i software with discipline-specific land development features and functionality. These include powerful and intuitive tools that let you create and label survey points, define and edit parcels and roadway alignments, automate drafting procedures, create terrain models, and calculate volumes and contours.

At the core is a centralized project structure that allows you to work more efficiently on any size or type of project. All of your critical data—points, terrain models, and alignments—is stored in a central location where it can be shared with others and used to create drawings. For example, 3D grids, contours, and sections are all extracted from a project's terrain model. When the project changes, your entire project team can quickly react—all working from the same data.

Exploring Your Design Options

AutoCAD Land Development Desktop 2i is perfect for experimenting with what-if scenarios: How does water flow across this site? What are the average slopes? What environmental responses may be needed? In just minutes you can combine existing drawing files (AutoCAD® DWG, MicroStation DGN, and others) with raster imagery, point data, and polygons from GIS sources (such as ESRI ARC/INFO coverages) and then build terrain models that display topographic conditions across the site. Not only do you have a drawing on which to explore design scenarios, but

---

[1] Autodesk, the Autodesk logo, AutoCAD, AutoCAD Map, and CAD Overlay are registered trademarks of Autodesk, Inc., in the USA and/or other countries. All other brand names, product names, or trademarks belong to their respective holders.

C.Y. Westort (Ed.): DEM 2001, LNCS 2181, pp. 39–42, 2001.
© Springer-Verlag Berlin Heidelberg 2001

you're also working with centralized project data that streamlines the process. Your projects can move more efficiently from the initial design phase all the way to final documentation.

This release incorporates the new features and performance of AutoCAD 2000i, the latest release of AutoCAD software. Internet-enabled design is one benefit, and so is a 24 percent productivity gain over the previous release. It also provides land development–specific functionality including DEM support and automated slope annotation. Upgrade today to gain a critical competitive advantage.

Product Features

AutoCAD® Land Development Desktop 2i software streamlines productivity for each phase of a project and provides a central location for project data that every professional on your team can access. It has tools for the entire design team and the entire design process.

Setting Up the Project and the Drawings

- Store and retrieve drawings and project data as part of a project.
- Use real-world coordinates and exaggerated vertical scale.

Using All the Functionality of AutoCAD Map® 2000i

- Access and compile data from numerous sources, including AutoCAD® DWG, MicroStation DGN, ESRI ARC/INFO coverages, and many others.
- Create drawings and rubbersheet, digitize, and clean up automatically.

Creating and Managing COGO Point Data

- Have complete control over appearance and visibility of points in drawings.
- Store a virtually unlimited number of COGO points in external project databases.
- Create and retrieve point groups based on various data criteria.
- Associate additional point data, such as borehole values or external databases.

Creating Base Geometry

- Create lines based on bearing, azimuth, turned angle, length, and more.
- Produce curves based on a variety of graphical and geometric input methods.
- Lay out clothoid, sinusoidal, cosinusoidal, and quadratic spirals.

Annotating

- Add multiple labels and tags to lines, curves, spirals, and polylines.
- Define labels with styles to maintain drawing integrity.
- Update label values, locations, and bearing direction automatically.
- Create multiple tables in one drawing.

Defining and Manipulating Alignments

- Define plan alignments that represent roads, rails, and channels.
- Edit graphically or with a tabular editor.
- Create offset geometry, station labels, and station offset values automatically.

Creating and Managing Parcels

- Size parcels with standard methodologies.
- Generate reports, including area, perimeter, and map check.
- Label parcels automatically.

Terrain Modeling and Analysis

- Create and manage surfaces with Terrain Model Explorer.
- Build surfaces from any combination of project data: point data, point groups, contours, AutoCAD objects, breaklines, ASCII point files, and DEM data.
- Access surface data without TIN objects added to the drawing.
- Use a variety of surface-editing commands.

Volumes

- Use composite, grid, and sectional methods to calculate volumes.
- Create volume reports, plot cross sections, and generate cut and/or fill contours.

Contours

- Generate contours at user-defined interval, layer, and style automatically.
- Control appearance of drawing's different contour groups with style-based system.
- Trim, extend, and edit contour objects.

Analyzing Terrain Conditions

- Update section baselines automatically.
- Extract and plot sections from multiple surfaces.
- Use thematic mapping for slope shading and elevation banding.
- Visualize 3D surfaces, including 3D grid and surface triangles.
- Define flow lines along surfaces.
- Calculate watershed areas and subareas.

System Requirements

- Intel Pentium–based PC with 90MHz processor
- Microsoft Windows 2000, Windows 98, Windows 95, or Windows NT 4.0
- 64MB RAM
- 800⨉600⨉64K display

Act Today!

Purchase AutoCAD Land Development Desktop 2i software at your preferred Authorized Autodesk® Reseller or Distributor or your Autodesk Systems Center (ASC). To find the nearest reseller, distributor, or ASC, phone or fax the appropriate number below.

United States and Canada
800-964-6432

Latin America
415-507-6110 fax

Asia Pacific
408-517-1748 fax

Europe, the Middle East, and Africa
+41-32-723-9394 fax
For more information, visit www.autodesk.com/landdesktop.

You can extend AutoCAD Land Development Desktop 2i's functionality by adding the following focused applications: Autodesk® Civil Design 2i, Autodesk® Survey 2i, and Autodesk® CAD Overlay® 2000i software. Visit the Autodesk website at www.autodesk.com for more information on these products.

Procedural Geometry for Real-Time Terrain Visualisation in Blueberry3D

Mattias Widmark

Sjol and & Thyselius Virtual Reality Systems
`mattias.widmark@blueberry3d.com`

Abstract. A common concern of the real-time terrain visualisation of today is the lack of geometric detail, when the terrain model is studied close enough. This is primary not a problem of the digital terrain data used being of too low resolution. The real reason is that the terrain model is pre-generated and stored, thus imposing database size and retrieval rate limitations to the model.

The present paper introduces an alternative technique, with which most geometric detail is created only when needed (*procedural geometry*). This renders many important benefits, including unlimited detail, no repetition of geometry, and optimal system resources usage. An implementation of the procedural geometry technique, Blueberry3D, is presented, illustrating examples of the kinds of geometric detail that allows itself to be procedurally generated.

1 Background

Most real-time terrain visualisation packages of today rely on a simple concept. An orthographic aerial photo or satellite image is draped upon a height model. The image can vary in spatial and colour resolution, and there are countless variations of the height model, but some important characteristics are common amongst all solutions.

The most serious concern is the limitations of the resolution of the raw map data. For elevation data, a resolution of tens of meters is common, and for imaginary data no better than meter resolution is usually available for larger areas. Resolution increases, and new schemes for handling very detailed elevation data are beeing developed[1] but the resolution will still be limited. At a certain distance from the virtual terrain, it will always turn flat looking and uninteresting.

One way to remedy this problem is to generate content, like foliage, for the terrain. This approach will decrease the minimum distance where the terrain looks rich and alive, but the problem remains. Pre-generated content needs to be bound in resolution due to storage space and retrieval speed limitations.

---

[1] The ROAM algorithm [ROAM] is a good example of an algorithm handling elevation data of very high resolutions by a clever LOD scheme.

C.Y. Westort (Ed.): DEM 2001, LNCS 2181, pp. 43–47, 2001.
© Springer-Verlag Berlin Heidelberg 2001

2   Procedural Geometry

2.1   Definition

Procedural Geometry refers to geometrical shapes created programmatically. For example, take a sphere-drawing routine. You supply it with a centre and a radius and the routine finds an appropriate set of triangles that together will, at least approximately, describe the surface of the sphere. These triangles are an instance of procedural geometry, and may later be usedfor the actual drawing of the sphere.

In this paper we adopt a more narrow definition. With procedural geometry, we only refer to shapes created *in real-time*, that is, shapes are created at the time they are needed, as opposed to shapes created by a pre-processing routine and stored for later. This rules out most existing terrain visualisation and related systems[2] that use procedural geometry from our discussion, since these systems only use the wider definition.

It is important to note that procedural geometry, when used in the context of this paper, mainly is focused on creating content on resolutions below those of the raw terrain data. At higher resolutions, pre-processed terrain data is used for the visualisation.

2.2   Benefits

Six main potential benefits from using procedural geometry for real-time terrain visualisation have been identified.

1. *Unlimited resolution.* Detail down to centimetre scale and below can be created applying fractals for computation of procedural geometry.
2. *Richness.* In practice, no repetition at all will occur, even for vast areas, by applying controlled randomness to the procedure.
3. *Natural terrain.* The unlimited resolution, fractals and random richness all serve to create natural areas that really come alive, much more than what can be achieved by traditional techniques.
4. *Decreased design time.* If made by hand, the small details in a terrain model are the most expensive to design. When procedural geometry is used, huge amounts of detail can be abstracted by small procedures using controlled random numbers.
5. *Compact databases.* The database size can be kept to a minimum by keeping most of the content stored implicitly in the program itself.
6. *Optimal system usage.* A terrain engine using procedural geometry can easily adopt the detail to comply with the hardware at hand, as there is no limit to the amount of detail available.

---

[2] Examples of programs using the wider concept of procedural geometry include *Bryce3D* [B3D], *Terragen* [TG], and *Onyx Tree* [OT].

2.3 Problems

Procedural geometry also comes with its share of downsides. Below, three issues are listed and addressed.

1. Procedural geometry is CPU intensive. However, even the standard pc:s of today are powerful enough to create rich terrains with procedural geometry [BB].
2. Procedural geometry demands a lot of system memory. This is mainly due to the fact that more detail than before can be created, but also that the generated shapes need to maintain extra state variables used by the generator. These variables would not be needed had the shapes been pre-created. In practice, 256 MB of memory is enough [BB].
3. Procedural geometry needs advanced level-of-detail handling. Not all the geometry of a terrain model can be generated and used at once, that would severely overload any hardware system. Rather, a scheme needs to be devised to decide where in the terrain model a high LOD is needed and where it is not. A related problem is called "plopping"; how do we change from one level of detail to another with as little visual interference as possible? The level-of-detail handling problem can be addressed in several ways, but there is not yet a perfect solution.

3 Procedural Geometry in the Blueberry3D System

Blueberry3D [BB] is a new terrain visualisation toolkit that uses standard formats for map data in combination with several different geometry generators. The result is an easy to use tool that, with little user effort, creates rich terrain models. In the following paragraphs, some of the geometry generators in Blueberry3D are presented as examples of the effects achievable with procedural geometry in the field of terrain visualisation.

3.1 Elevation

Ground elevation geometry is generated as a variation of two-dimensional *Fractal Brownian Motion (FBM)* [FR]. The "coarseness" of the surface is specified per *terrain class*[3].

3.2 Ground Structure

To achieve a realistic looking ground structure, several *ground layers* are used. A Blueberry3D ground layer usually maps to a real ground layer like soil, rock or sand. The layer is fitted with a set of parameters describing properties like erosion, coarseness and fertility. Each terrain class can have its own set of ground layers.

---

[3] A terrain class is a classification of some terrain, like boreal forest, grass plains and open rock.

The ground structure geometry generator takes all the local ground layers and their parameters into account when the geometry is generated. Many natural effects like rocks poking out of the surrounding soil (Fig. 1) and sea waves flushing away grass and dirt, leaving only sand, are obtainable by properly configured ground layers.

Fig. 1. Some procedural rocks poking out of the sand.

3.3   Vegetation Distribution

The distribution of vegetation items is performed by another variation of the FBM. The effect is that natural-looking glades and clusters of trees and bushes are created automatically. The fertility of the local ground is also taken into account.

3.4   Vegetation Items

Vegetation items, like individual trees and bushes, are modelled using *Iterated Function Systems (IFS)* [FR]. A species is created using the Blueberry3D interactive

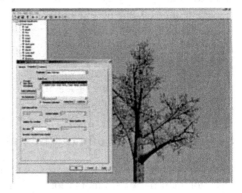

Fig. 2. Development version of Blueberry3D Interactive Tree Editor.

Tree Editor (Fig. 2), and unique instances are generated by the corresponding geometry generator and controlled random numbers.

3.5   Curved Surfaces

Blueberry3D uses polynomial surfaces to model roads, rivers, walls, paths, shafts, house foundations, etc. The main idea in all cases is that the user provides a minimum of information and the system takes care of the details of the modelling. For instance, a road is modelled as a set of points in the terrain and a profile. The curved surfaces generator interpolates the points, adopts the terrain around the road to make it blend nicely[4] and finally places the road (Fig. 3).

Curved surfaces, like all other procedural geometry, are computed in real-time. The result of this is that, no matter at how close range the user observes the surface, it will still look smooth. Only a few real-time systems have this functionality, one example (not terrain-visualisation related) is the OpenGL Optimizer from SGI [OO].

---

[4] This involves lowering, raising and smoothing of the terrain, removal of unwanted trees and bushes and possibly changing the surface texture.

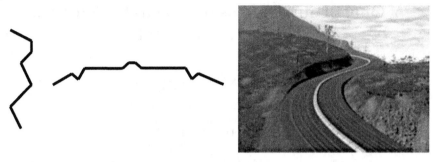

Fig. 3. Road as a point sequence in the map (left), a profile (middle) and the final result (right).

3.6   Conclusion

The Blueberry3D software shows that real-time terrain visualisation using procedural geometry is indeed achievable on a standard home computer, with all the benefits including increased detail and variation, huge terrain models and short design time. The technology has a great potential and it is our opinion that it will play an increasingly important role in the field of real-time terrain visualisation.

References

[B3D]   Corel Inc, *Boyce 4 software*, http://www.corel.com
[BB]   Sjoland & Thyselius VRS: *Blueberry3D software*,
          http://www.blueberrv3d.com
[FR]   Peitgen, H-O. & Saupe, P. (eds.), 1988: *The Science of Fractal Images*.
          Springer Verlag.
[OO]   Silicon Graphics: *OpenGL Optimizer Technical Info*,
          http://www.sgi.corn/software/optimizer/tech_info.html
[OT]   Onyx Computing: *Tree Software*, http://www.onvxtree.com
[ROAM]   M. Duchaineau et. al. 1997: *ROAMing Terrain: Real-time optimally adapting meshes*. Proc. Visualization '97, pages 81–88.
[TG]   Planetside Software: *Terragen Software*,
          http://www.planetside.co.uk

A Quasi-Four Dimensional Database for the Built Environment

Henning Sten Hansen

National Environmental Research Institute,
Ministry of Environment & Energy
Frederiksborgvej 399, DK-4000, Roskilde, Denmark
Telephone : +45 46 30 18 07
Fax : +45 46 30 12 12
HSH@DMU.DK

Abstract. The growth of cities represents huge problems for modern societies. The explanation of the phenomena involves the description true 3D space as well as changes over time. However, there is a general lack of true 4D data. The current paper describes how to combine generally available spatial databases and administrative registers to create a quasi-4D database for the built environment. Although this 4D database might seem simple, it is used widely in urban planning and environmental monitoring and modelling.

1 Introduction

The problems related to the growth of cities and the concentration of human population into large metropolitan areas represent huge challenges for modern societies. Economic growth drives urban expansion in the form of construction of businesses, dwellings, roads, leisure centres etc., and the metropolitan regions face the growing problems of urban sprawl, including a decline in natural vegetation, wildlife habitats and agricultural land. Thus the replacement of undeveloped land by residential and commercial development continues at an unprecedented rate.

The study of geography is basically 2-dimensional. However, in order to understand many environmental and planning phenomena it is often necessary to include the third dimension and to study how these patterns change over time. Architects and planners have for more than ten years used 3D based CAD systems for design and planning of new urban landscapes. Furthermore, geographers and cartographers have long attempted to visualise changes by integrating spatial and temporal information on a series of maps [1]. Unfortunately, only a few attempts have been done on a true integration of 3D space and time [8]. The background for this situation is primarily a general lack of spatio-temporal data. In addition, no current available GIS software includes a full 4D data model, although a few attempts have been done.

During the last 25 years a lot of nationwide databases with spatial information are established, and in the late nineties the nationwide Cadastral Map and

C.Y. Westort (Ed.): DEM 2001, LNCS 2181, pp. 48–59, 2001.

the national topographic map - TOP10DK - was produced. However, real 4D (spatio-temporal) data are generally not available, and we have to combine and manipulate existing data sources to produce quasi-4D databases. The purpose of the current project has been to produce a quasi 4D database for buildings in Danish urban areas based on generally available spatial databases with a nation wide coverage. A database like this has been requested heavily among urban planners and environmental scientists for a long time.

The current paper is divided into five main parts. First, I present some basic concepts of space and time and how these concepts can be applied within the current study. Next, the basic data sources are discussed. Third, I describe how to build a quasi spatio-temporal database covering the built environment. Fourth, I describe how the quasi spatio-temporal database can be used in physical planning and environmental monitoring and analysis. Finally, a few concluding remarks and a presentation of subsequent research.

2   Space and Time

While the third dimension is an obvious extension of the 2D space, time is a bit more difficult to handle conceptually. You cannot imagine a coordinate system with four axes! However, there are no fundamental difficulties handling 4D data compared to 2D data. Usually, time is assumed to be linear and similar to space. Therefore, time can be considered as the fourth dimension in addition to the three-dimensional space. The most general way of storing temporally referenced real world data is in the form of four-dimensional (4D) objects in a 4D time-space. Therefore, a spatio-temporal GIS should incorporate time as a fundamental component of the database. In addition to a traditional GIS, a spatio-temporal GIS is able to search for temporal patterns. Thus, the user can search applying both the "where" and "when" clause.

Time in a spatio-temporal information database may be measured as a discrete or continuous variable [14]. While a continuous variable assumes that values at any time can be obtained by interpolation, a discrete temporal variable means that the variation is discontinuous between the time of measurements. Obviously, urban growth phenomena belong to the last category.

The information to provide a true spatio-temporal database for the urban landscape ten or fifty years ago is not available. Traditionally cartographers have used map series of different time periods to get an idea of former states of the world. This so-called time-slice approach is an intuitively appealing model, and is often the only possibility to provide information on landscape changes [9]. However, the snapshots represent states - the situation at the times of map creation - rather than changes, providing no information on how and when the changes have taken place. The various map series are usually many years apart causing a very low temporal resolution.

Simple cartographic snapshots are not sufficient for highly dynamic phenomena. Time must be incorporated as a fundamental component similar to the geometry. For movements in 3D space you need simultaneous recordings of x, y, z and t at a rather high temporal resolution - today frequently obtained by GPS techniques. Alternatively, the geometry can be 3D with an associated time attribute or 2D with height and time attributes. The latter model is used in the current project - therefore the term quasi-4D.

3   Data Sources

Traditional spatial information systems retain only the latest state of the modelled system, presenting an up-to-date, but static view of the world. Furthermore, there is a general lack of data with a time dimension as well as three space dimensions. Nevertheless, some data sources can be combined to provide the needed information to create spatio-temporal data.

The current study is based on three national data sets: the National topographic Map database (TOP10DK), the Digital cadastral Map database, the Parcel Register and the Danish Building & Dwelling Register (BBR in Danish). A higher level of geometric accuracy might be obtained if TOP10DK is replaced by high quality technical maps produced by some municipalities. Unfortunately, these maps are of varying quality and not general available for all municipalities.

3.1   The Digital Topographic Map

TOP10DK is a national digital topographic base map, where the basic information regarding land-scape, municipal boundaries, names etc. are stored. TOP10DK - an acronym for **Top**ographic database with contents and accuracy corresponding to a similar analogue map on the scale 1 : **10**.000 covering the whole of **D**enmark [2]. The geometry is based on new photogrammetric registrations and existing municipal / utility map databases. For well defined objects, the relative and absolute accuracy is better than 1 metre in all three axes. TOP10DK are available for the whole country from the end of year 2000. Basically TOP10DK contains <u>no</u> foreign database keys. In the current study only the buildings in TOP10DK are used.

3.2   The Digital Cadastral Map

The Danish National Survey and Cadastre administrates real property data in Denmark. This involves the digital cadastral boundary map as well as the Parcel Register and the Building and Dwelling Register.

The digital cadastral boundary map displays the existing cadastral structure. The basic unit within the cadastral map database is the individual parcel. The database includes cadastral boundaries, parcel identifiers, names of municipalities, parishes etc. [10]. The accuracy of the dataset corresponds to the scale of the original analogue maps, i.e. scale 1 : 4000. The principal function of the cadastre is the maintenance of an up-to-date database of all land parcels in Denmark. The digital cadastral map is continually updated as cadastral surveys are completed. The dataset covers the entire country. A composite key defined by the *cadastral district number* and the *cadastral number* acts as the primary key in the cadastral map database. Additional information is stored as attributes, such as property number, property owner, mode of exploitation, etc. For historical reasons, these data are stored in a separate parcel database. The linkage between the two databases is based on a composite key defined by the cadastral district number and the cadastral number. These numbers are assigned to each parcel.

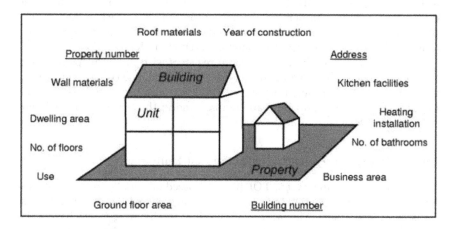

Fig. 1. The Building & Dwelling Register – entities and attributes.

3.3 The Building & Dwelling Register

The Building and Dwelling Register managed by the Ministry of Housing was established in 1977 and contains detailed information on all buildings in Denmark [11]. This register uses 3 levels of registration (figure 1).

- Property level - type of ownership, sewage disposal system etc. are stored. A *unique property* number is the primary key at this level.
- Building level - purpose for which the building is used, year of construction, roof material, number of floors etc. are stored. The individual buildings are identified by a composite key made by a *property number* and a *building number*.
- Unit level - area of the unit, number of rooms, kitchen facilities etc. are stored. The dwelling units are identified by *addresses*.

Besides this, the Building & Dwelling Register contains some additional foreign keys (e.g. municipality number, road code and house number / letter). In the current context, the number of floors and the year of construction play a key role, making possible the assignment of a vertical as well as a temporal dimension to the data. The temporal granularity in this case is one year, although the Building & Dwelling Register is updated continuously.

4   Creating the Database

Using the National topographic map, the Digital cadastral Map and the Building & Dwelling Register a 4D database for the built environment is created. The digital topographic map delivers the basic geometric information - the ground plan of the buildings and eventually the terrain surface. The third and temporal dimensions are obtained from the Building & Dwelling Register.

Linking data stored in administrative registers with digital maps require a unique relationship between the objects in the map and the corresponding rows in the register. This means that object definitions and database keys must be exactly the same in the digital map database and in the corresponding administrative register.

4.1   Defining the Footprint of Individual Buildings

The national topographic map, TOP10DK, is based on aerial photos, and therefore the buildings in densely built-up areas are aggregated into building blocks. Thus, in order to define the individual buildings the building blocks have to be slit along parcel boundaries and finally assigned appropriate database keys. This process involves the following steps:

- Generation of a property map. The property is the main part of a composite primary key in the Building and Dwelling Register. A property map is generated, by spatial merging cadastral polygons with the same property number.
- Unsplitting multi – part property polygons. The property map contains some multi-part polygons; these are unsplit into single-part polygons. The resulting property objects correspond to the records in the Building and Dwelling Register at the property level (figure 2).
- Forming individual building polygons by overlaying. Comparing the building theme and the property theme indicates the need to divide up some building blocks into individual buildings. Overlaying the original building theme with the property theme will split the building blocks along the property boundaries, thus creating a new theme containing the individual buildings. Each building unit now contains a property number - the main part of the composite primary key in the Building and Dwelling Register.
- Removing slivers. Intersecting one map theme with another may cause slivers to be generated due to minor differences in the map production specifications and techniques. These slivers are removed.
- Geocoding buildings by area matching. The final step in this process concerns the assignment of a unique key to every building. Assigning building

numbers to the buildings is quite complicated, but comparing the ground floor area of each building in the topographic map and the Building & Dwelling Register has appeared to be a useful approach.

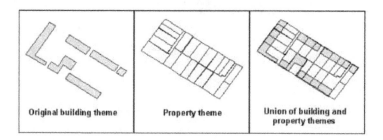

Fig. 2. Cutting building blocks into individual buildings.

The result of this process - described more detailed in Hansen [5] - is a new map theme containing the footprint of each individual building. Furthermore, most buildings are assigned a property number and a building number. Due to errors in the digital maps as well as the Building & Dwelling Register some buildings (about 10 %) are not assigned property number and building number. Consequently, these buildings cannot be assigned additional information from the Building & Dwelling Register.

4.2 Assigning Height Information to the Buildings

The third dimension is equal to the height of each building. The Building & Dwelling Register does not include explicit information concerning building heights. However, the height of the building can be estimated based on the following building characteristics: number of floors (3 meter per floor) and type of roof (pitched is set to 2.5 meter and flat to 0.5 meter). Thus, a generalised 3D model of each building can be made (figure 3). Evidently, this generalisation might seem too coarse, and a more detailed 3D model might be obtained by photo-grammetric methods [3], but for overall visualisation and modelling purposes some kind of generalisation are preferable. Too many details will slow down the calculations. In practise it might be difficult precisely to visualise the true roof construction. We have tried to make simple models of the roofs, but it is too time-consuming - too much manual work - to add "true" roof constructions on every building.

4.3 Adding a Temporal Dimension to the Database

In this respect, the year of construction plays a key role, making possible the assignment of a temporal dimension to the data. The temporal granularity in

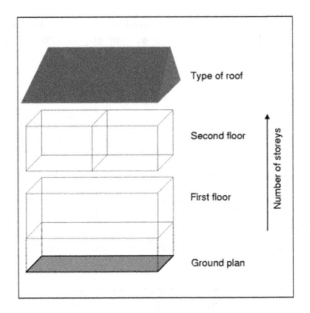

Fig. 3. Conceptual model of the buildings

this case is one year, although the Building & Dwelling Register is being updated continuously by the municipalities. Copies of the Building & Dwelling Register have been available for more than twenty years. Furthermore, the Building & Dwelling Register contains a database with all changes since 1977. However, the historic database might be difficult to use, due to changing database keys, lack of old digital maps etc.

The kind of spatio-temporal data model adopted here, was originally developed by Peuquet and Duan [12]. This so-called event-based spatio-temporal data model implies a concept of some base state with subsequent amendments - represented by new, changed or demolished buildings (figure 4). Currently, a copy of the Building & Dwelling Register for only one single year - 1998 - is used. The main problem introduced by this approach is the missing possibility to track the buildings through their whole life cycle from creation to demolition - only the creation is considered. However, regarding only shorter time periods - less than fifty years - the errors will be minimized. In this case the problems are restricted to core urban areas with large-scale urban renewal projects. Therefore, you have to be careful, if you observe new buildings in densely built-up areas. Generally, this will represent a renewal project.

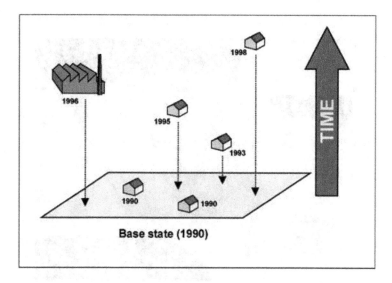

Fig. 4. The temporal dimension of the building database.

5   Examples

The produced quasi-4D database for buildings has a wide range of possible applications within physical planning and environmental monitoring and modelling.
First, I will present some general extraction from the database. Figure 5 illustrates a 3D view of the urban land-use in a minor part of the Danish town
Middelfart located in the westernmost part of the island Funen. The 3D view
- representing the year 1998 - shows four different land-use categories obtained
from land-use code in the Building & Dwelling Register. Compared with traditional two-dimensional land-use maps, a 3D view gives a more detailed presentation
including the density of the built-up area

The next example - figure 6 - shows buildings in 3D for two different years
- in this case a part of Kirke Stillinge, Western Zealand. All buildings in the
figure are detached houses, and 1970 represents the beginning of an era of urban
expansion - even in smaller villages like Kirke Stillinge. The ground surface in
figure 6 represents a digital terrain model (TIN surface) generated from a contour
map.

Both examples are made using ArcView 3D Analyst. The main disadvantage
concerning 3D Analyst is its very simple way to handle real 3D objects. The
buildings in the figures below are just extruded and placed on top of the terrain
model. However, the simple visualisation tools in 3D Analyst fits quite well the
quasi-4D building database, even the buildings are expanded with roofs.

The 4D database described above have been used in a lot of research projects
within our National Environmental Research Institute. First, it was used as a
representation of the urban landscape in detailed street pollution modelling [4],

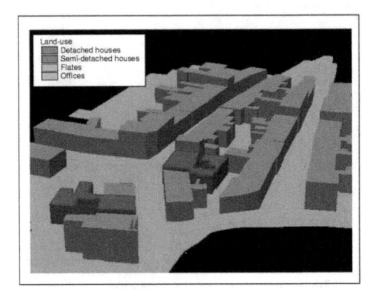

Fig. 5. 3D view of urban land-use in a minor part of Middelfart.

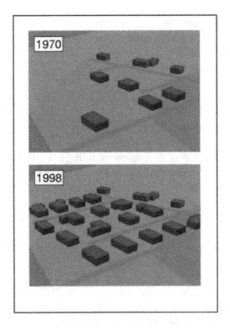

Fig. 6. 3D scenes for various years – the village Kirke Stillinge, western Zealand.

[7]. Re-circulation of air in a street canyon with the wind perpendicular to the street orientation gives higher concentrations of pollution at the leeward side than at the windward side. Consequently, a 3D (or $2^1/_2$D) urban landscape model has to be applied as input to the so-called Operational Street Pollution Model developed by NERI.

Next, it was used in an analysis of urban growth in Copenhagen Metropolitan area since the end of the Second World War in 1945 [6]. Using the spatio-temporal temporal database, animation frames were created for each year in the period 1945 - 1995, and based on these frames animated GIF89a files and AVI files were created. The animated GIF is very useful for visualising dynamic maps on the Internet, thus increasing the ability to spread the information to a wide audience. The so-called Aarhus Convention [13] ensures access to environmental information and public participation in the decision making process concerning environmental problems and in this context, dynamic visualisation of spatial data will be of great importance

6   Concluding Remarks

The globalisation drives economic growth and urban expansion in the form of construction and the metropolitan regions face the growing problems of urban sprawl, including a decline in natural vegetation, wildlife habitats and agricultural land. Thus, there is a need for still more sophisticated analyses and models in urban planning and environmental science in order to keep the growth sustainable. Spatial information systems with 3D geometry as well as temporality seem to be an important tool in this work. However, there is a general lack of 4D data. Therefore, the current paper has described how to use existing map databases and administrative public registers to create a quasi- four-dimensional database for the built environment.

Since 1977, the Danish municipalities have stored detailed information about every single building in Denmark. This so-called Building & Dwelling Register includes among other data building use, year of construction and number of floors. Additionally, Denmark is covered by high quality digital topographic and cadastral maps. Combining these data sources is not an easy task due to lack of correspondence between the building entities in the topographic map and the rows of the Building & Dwelling Register. Furthermore, the buildings in the topographic map do not contain any database key. Automatic methods to do this task is developed during the current project and the building entities are joined with the Building & Dwelling Register.

The result is new database with 2D geometry and building height (third dimension) and year of construction (fourth dimension) as attributes. Using the number of floor as an indicator of building height and the year of construction as the temporal dimension do not fulfil the requirements of a true 4D spatio-temporal database.

However, the created database for the built environment is very useful in practical planning and research where the vertical dimension and time is important components. As mentioned above, this database has already been used in air pollution modelling and urban growth studies.

Nevertheless, the 4D database can be improved. First, we will try to use the historical part of the Building & Dwelling Register in order to handle demolition of buildings (urban renewal) in a more proper way. This might improve the possibilities of tracking urban land-use changes for tens of years. Second, we will try to improve the vertical dimension in order to take into consideration vertical changes in land-use - i.e. land-use classification based on units instead of buildings.

Acknowledgements. I would like to acknowledge the financial support from the Ministry of Environment and Energy and helpful suggestions during the course of this research from Steen Solvang Jensen (NERI) and Hans Skov-Petersen (National Forest and Landscape Research Institute).

References

1. Acevedo, W. & Masuoka, P. (1997). Time-series animation techniques for visualizing urban growth. *Computers & Geosciences*, vol. 23, pp. 423–435.
2. Eggers, O. (1996). TOP10DK - Danish solution to digital topographical mapping. In: M. Rumor,., R. McMillan & Ottens, H.F.L. (eds.): *Geographical Information. From Research to Application through Cooperation.* 2 nd Joint European Conference & Exhibition on Geographical Information. Barcelona, Spain, 1996. IOS Press. pp. 999–1002.
3. Förstner, W. (1999). 3D-City Models: Automatic and Semiautomatic Acquisition Methods. In Fritsch, D. & Spiller, R. (eds): *Photogrammetric Week'99*, Wichmann, Karlsruhe.
4. Hansen, H.S., Jensen, S.S. & Berkowicz, R. (1997): Estimating Street Air Quality Using a $2^1/_2$ Dimensional Urban Landscape Model. Den *Store Nordiske GIS-Konference*. Kolding 29.-31. oktober 1997. DAiSI, DKS, DSFL, AM/FM-GIS Nordic Region.
5. Hansen, H.S. (1999). Integrating digital maps and administrative registers - Danish experiences. 21 st *Urban Data Management Symposium*, 21 - 23 april 1999, Venezia.
6. Hansen, H.S. (2001). A time-series animation of urban growth in Copenhagen metropolitan Area. In Bjørke, J.T. & Tveite, H. (eds.): *Proceedings 8[t]h Scandinavian Research Conference on Geographical Information Science*, Ås, Norway, pp. 225–235.
7. Jensen, S.S., Berkowicz, R., Hansen, H.S. & Hertel, O. (2001) A Danish decision-support GIS tool for management of urban air quality and human exposures. *Journal of Transportation Research.* Part D Transport and Environment, vol. 6, pp. 229–241.
8. Könninger, A. & Bartel, S. (1998). 3D-GIS for urban purposes. *GeoInformatica*, vol. 2, pp. 79–103.
9. Langran, G. (1992). *Time in Geographic Information Systems.* Taylor & Francis, London.
10. National Survey and Cadastre (1997). *The Digital Cadastral Map - User Guide.* National Survey and Cadastre, Copenhagen, 1997. (in Danish)

11. National Survey & Cadastre (1999). *User Guide for the Building & Dwelling Register.* National Survey and Cadastre, Copenhagen, 1999. (in Danish)
12. Peuquet, D.J. & Duan, N. (1995). An event-based spatio-temporal data model (ESTDM) for temporal analysis of geographical data. *Int. J. Geographical Information Systems*, vol. 9, pp. 7–24.
13. UNECE (1998). The Aarhus Convention.
    `www.unece.org/env/pp/contentofaarhus.htm`
14. Worboys, M.F. (1995). GIS: A *Computing Perspective.* Taylor & Francis, London.

Interactive Generation of Digital Terrain Models Using Multiple Data Sources

Angelika Weber[1] and Joachim Benner[2]

Forschungszentrum Karlsruhe, Institut für Angewandte Informatik,
Postfach 3640, D-76021 Karlsruhe, Germany
[1] weber@iai.fzk.de
[2] benner@iai.fzk.de

Abstract. The work presented aims to develop an adaptive digital terrain model, represented as TIN, by combining different kinds of input data like 3D data-grids, contour lines, and single measurement points. A specialized triangulation process is being developed to optimally integrate the data sets and to estimate an overall error. Special emphasis is laid on the generation and integration of polygons extracted from scanned topographical maps. The terrain modeller is part of the MELINDA information system for waste dumps and contaminated sites.

Keywords: Terrain Model, TIN, Model Accuracy, Triangulation, Scanned Maps.

1 Introduction

A central problem with digital terrain models is the conflict between model accuracy on the one hand and model size and cost on the other hand. Normally, highly accurate models are very large and difficult to handle with standard PC-systems, and the cost of input data is high. Our approach therefore is to produce adaptive terrain models, with high accuracy only at specific "areas of interest", and lower accuracy elsewhere. This enables the usage of freely available digital elevation data (like e.g. the USGS GTOPO 30 model [1], or data extracted from topographic maps), locally modified and refined with other data, e.g. from surveying missions or land-register authorities.

Here "accuracy" means a measure for the discrepancy between the model, represented by a Triangulated Irregular Network (TIN), and the real terrain. This measure has to take into account measurement uncertainties, as well as approximation errors of the triangulation surface. Algorithms have to be developed estimating these errors for a specific input data set, as a function of space, and estimating the error propagation when different data sets are combined. A corresponding methodology is described in [2], [3].

The terrain modeller being developed will furthermore support the extraction of geo-referenced contour lines and topographic lines from scanned maps, offering a flexible and easily available source for digital terrain data. It is a sub-module of the MELINDA information system for waste dumps and contaminated sites [4], [5]. MELINDA is designed to manage all kinds of data available for a certain site,

C.Y. Westort (Ed.): DEM 2001, LNCS 2181, pp. 60-64, 2001.

including geographical, geological, hydrological, chemical, and CAD-information. The information system is able to visualize the 3D data in a common coordinate system, with user-defined choice of viewing position, viewing direction, and projection method. The user has geographic access to all kinds of information and is able to "fly" through the whole site in VR-manner. MELINDA also provides a number of interactive editors for model generation and modification. Among these is a special TIN-editor, used for manually adapting the TIN-geometry.

2   Data Sources

In general, three different kinds of data sources can be identified for the creation of a 3D terrain model.

- 3D data-grids, produced from satellite or airplane images with photogrammetric methods. Here, the measurement-uncertainties normally are known, and data cost increases with accuracy. This kind of data can be handled with a number of GIS or CAD-systems like ArcView/3D Analyst or AutoCAD.
- Digital and analog surveying data, which either already exist, (e.g. at a local authority) or are produced during a surveying mission. In this case, data accuracy strongly depends on the chosen measurement method.
- Topographic maps with contour lines and topographic lines (like edges, roads, rivers, railway-lines, ...). Maps are easily available and not expensive, but they are difficult and time-consuming to process with standard digitising hardware (tablet) and software. It is more efficient to extract lines directly from a scanned map, with user interaction to support missing data as, e.g. the height-attribute of a contour line or the vertical (Z-) coordinates of a topographic line.

Integrating different data sources results in an unsorted collection of 3D grids, contour lines, 3D topographic lines, and 3D points with varying accuracy, eventually referring to different reference systems or ellipsoids. While the transformation into a common reference system is a standard task, the integration of data with different accuracy levels is much more complex and not yet solved sufficiently.

3   Realisation

3.1 Realised Functionality in MELINDA

Up to now, modules and algorithms have been developed to generate TINs on basis of single data sources (figure 1). An incremental Delaunay triangulation algorithm [7] has been realised, working on regular or irregular grids of 3D points. Topographic lines can be regarded as restrictions. The system is able to produce triangulation surfaces with a minimal number of triangles for a given approximation error. This error is defined as the maximal vertical distance to the triangulation surface of all control points not regarded during the incremental triangulation process.

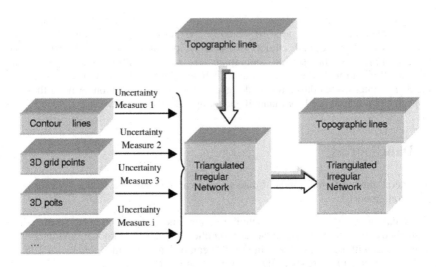

Fig. 1. Triangulation process

Furthermore, it is possible to generate a TIN from a collection of contour lines and topographic lines. Again, the algorithm tries to minimize the number of triangles while regarding a prescribed approximation error.

Functions for user defined modification of TINs also exist. The TIN geometry can be modified by projecting 3D topographic lines or profile surfaces into the triangulation surface. A profile surface, defined by sweeping a profile polygon along a topographic line, can be regarded as simple geometric model of e.g. a river valley. Geometrical modifications are also possible on basis of single nodes, edges or triangles in the TIN.

For rendering a TIN, the user can freely define viewing position and angle, camera projection method, and lighting. The TIN can be coloured with respect to the vertical height, using an arbitrary colour table. Alternatively, a geo-referenced raster-image can be projected onto the triangle surface (figure 2). Data import and export supports different interface formats, including VRML-model generation.

3.2 Map-Editor and Data Fusion

A 2D-editor for identifying, drawing and modifying different kinds of lines in a scanned map is under development. We distinguish between contour lines and topographic lines, where the latter ones are polygons representing topographic objects like breaklines, rivers, edges, etc. By user interaction, every line gets additional attributes like height-attribute, colour, profile polygon, and an estimated uncertainty measure.

Fig. 2. TIN-model with projected orthophotos

From this data set, only the contour lines are used in the first step of the triangulation process. If available, 3D-grid data and data sets with single measurement points can be used in addition. The triangulation algorithm combines all data sets to generate a first triangulation surface. In the next step, this surface is modified by integrating topographic lines. This is performed by projection algorithms specific for the topographic object corresponding to the line. The projection method for a river, e.g. will enforce an always negative gradient of the corresponding line.

4   Summary and Outlook

The MELINDA terrain modeller as flexible tool for generating Digital Elevation Models has been presented. Central components are a comfortable map editor for extraction of contour lines and topographic lines from scanned maps and a flexible triangulation algorithm integrating grid, line and point data. Though the system is being developed in the context of an information system for contaminated sites, a lot of other application areas exist. High quality terrain models could be used in the construction planning process of new roads, parks and residential quarters. In the tourist area, trails for walking or cycling could be planned or presented in a 3D environment. VRML-models generated with the terrain modeller could be used as input for VR-tools or rendering and animation software, offering a wide range of applications in the advertising or computer game industry.

References

1. USGS EROS Data Center: http://edcdaac.usgs.gov/gtopo30/gtopo30.html.
2. M. Glemser, U. Klein, D. Fritsch: Complex Analysis Methods in Hybrid GIS Using Uncertain Data, GIS – Geo-Informationssysteme 2/2000, pp. 34 – 40 (2000).
3. M. Glemser, U. Klein: Hybrid Modelling and Analysis of Uncertain Data, IARPS, Vol XXXIII, Amsterdam, Part B4/2, Comm IV, pp. 491 – 498 (2000).
4. J. Benner, K. Leinemann, A. Ludwig, A. Weber: MELINDA – Ein multimediales Leit- und Informationssystem für Deponien und Altlasten, in : C. Rautenstrauch, M. Schenk (Ed.): Umweltinformatik 99, 13. Int. Symposium „Informatik für den Umweltschutz" der Gesellschaft für Informatik, Magdeburg (1999), pp. 206 – 219.
5. J. Benner, K. Leinemann, A. Ludwig, A. Weber: MELINDA – Interaktives Altlasten Monitoring- und Bewertungssystem; Proc. AGIT 2001 Symposium, Salzburg, 4. – 6. 7. 2001.
6. M. Garland, P. S. Heckbert: Fast Polygonal Approximation of Terrains and Height Fields, http://www.uiuc.edu/~garland/software/scape.html

The Shuttle Radar Topography Mission

Thomas A. Hennig[1], Jeffrey L. Kretsch[2], Charles J. Pessagno[3],
Paul H. Salamonowicz[2], and William L. Stein[2]

[1] National Imagery & Mapping Agency
Directorate of Operations
4600 Sangamore Road
Bethesda, Maryland, 20816-5003, USA
[2] National Imagery & Mapping Agency
Technology Office
12300 Sunrise Valley Drive
Reston, Virginia, 22091, USA
{hennigta, kretschj, pessagnoc, salamonp, steinb}@nima.mil

Abstract. Elevation data is vital to successful mission planning, operations and readiness. Traditional methods for producing elevation data are very expensive and time consuming; major cloud belts would never be completed with existing methods. The Shuttle Radar Topography Mission (SRTM) was selected in 1995 as the best means of supplying nearly global, accurate elevation data. The SRTM is an interferometric SAR system that flew during 11-22 February 2000 aboard NASA's Space Shuttle Endeavour and collected highly specialized data that will allow the generation of Digital Terrain Elevation Data Level 2 (DTED® 2). The result of the SRTM will increase the United States Government's coverage of vital and detailed DTED® 2 from less than 5% to 80% of the Earth's landmass. This paper describes the shuttle mission and its deliverables.

1   Introduction

The attempt to create a digital dataset called Digital Terrain Elevation Data (DTED®) has been ongoing since the 1970s. However, progress has been very slow, with about 70% of the world collected to DTED® 1 (3 arc second post spacing) standards, and less than 5% to DTED® 2 (1 arc second post spacing) standards. DTED® consists of a matrix of elevation posts at uniform spacing and specified elevation accuracy sampled to represent the earth's surface. It can be used to make elevation models and fly-thrus for mission planning, and modeling and simulation. Other examples of use include trafficability determinations, line of sight projections, determination of drainage, and use for route planning and airline safety.

The National Imagery and Mapping Agency (NIMA) is responsible for providing global imagery and geospatial information to government users. Elevation data provides a key ingredient to the readiness of NIMA's customers. Elevation

C.Y. Westort (Ed.): DEM 2001, LNCS 2181, pp. 65–77, 2001.
© Springer-Verlag Berlin Heidelberg 2001

data is one of three components of Foundation Data, which is comprised of elevation data, feature data, and imagery data. Both the Department of Defense (DoD) and the Intelligence Community have recognized the need for a global dataset of elevation at one arcsecond spacing since at least 1995.

Growing computer processing, display and communication capabilities made the need for DTED® support acute. In 1995, the DoD Joint Staff and the Intelligence Community recognized the need for a common understanding of the operational environment and that a major inhibitor was the lack of availability of digital terrain elevation data which is vital to successful mission planning, operations and readiness. The question was how to provide it quickly. For years the most efficient method to collect elevation data was stereo photogrammetric compilation. It was far more efficient than ground surveys, requiring only a few control points on the ground. A cartographer at an analytical plotter could map vast tracks of land. However, the work is labor intensive. Furthermore, much of the world is almost perpetually cloud covered. This, combined with competition for the imaging resources for purposes other than mapping, made acquisition of adequate source over much of the globe impractical. Several converging developments were to lead to a solution to the problem.

2   Interferometric Synthetic Aperture Radar

Synthetic Aperture Radar (SAR) was developed in the 1950s after being conceived by C.A Wiley. In the 1960's airborne imaging SAR systems were in operation. In the 1970s SEASAT, a spaceborne radar, operated for several weeks and demonstrated the possibilities of radar mapping from space. Use of SAR imagery addressed the cloud issue. However, most SAR systems are mono imaging systems. Stereo acquisition is a major collection issue and processing of stereo SAR imagery is labor intensive like the traditional photogrammetric approach. An overview of the history of SAR can be found in [1] and [2].

The development of the interferometric SAR (IFSAR) technique for obtaining elevation data was the last key item. A brief summary is provided here. For a more thorough presentation on the technique see [3] and [4]. The single pass IFSAR technique is based on using interference of the radar return at two separate antennas. A repeat pass approach can be used but suffers from major decorrelation issues. By the early 1990's an airborne IFSAR system was operational. Figure 1 illustrates the geometry of single pass IFSAR. A radar observation is made of the topography $Z(y)$ from each endpoint of the baseline B. In repeat pass interferometry these could be done at different times, in single pass they are done simultaneously. Several configurations can be used for single pass interferometry. The Shuttle Radar Topography Mission (SRTM) works by radiating a pulse from one antenna and measuring the difference in phase of the signal returns to each antenna. From Figure 1 we can derive the expression

$$Z(y) = h\{\lambda\phi/2\pi)^2 - B^2]/2(B\sin(\alpha - \theta) - (\lambda\phi/2\pi))\}\cos\theta \qquad (1)$$

where ϕ is the radar phase, λ is the wavelength of the radar, ρ is the range from the sensor to the ground, θ is the look angle, and h is the ellipsoid height. A process called phase unwrapping is required to determine the integer number of wavelengths to obtain the absolute range. An unwrapped elevation can then be calculated from Equation 1. A great deal more goes into generating digital elevation models, as later sections will indicate. The principal error source for uncertainty in height, σ_z is related to the uncertainty in the orientation angle, σ_α, by

$$\sigma_z = \rho \sin(\theta)\sigma_\alpha \tag{2}$$

Other error sources include errors in measurement of the radar phase of the return signals, the length of the baseline, and many other related measurements.

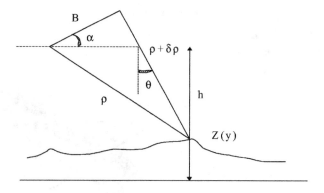

Fig. 1. Interferometric SAR geometry

3 The SRTM Mission

NIMA in concert with the National Aeronautics and Space Administration (NASA) and the Jet Propulsion Laboratory (JPL) sponsored the SRTM. Ed Caro and Dr Mike Kobrick of JPL conceived the SRTM essentially as a follow-on to the Shuttle Radar Laboratory (SRL) missions of 1994, using much of the SRL equipment modified to be an IFSAR system [6]. The inherited SRL hardware included the main C- and X-band antennas and the radar support structure. The key equipment added included a second set of antennas attached to the end of a deployable 60 meter mast, and a variety of specialized electronics systems. Figure 2 shows the Shuttle with the radar mast deployed in mapping configuration. Germany and Italy provided an X band IFSAR system called X-SAR. The C band system was called C-RADAR.

The SRTM orbit was inclined 57 degrees, the maximum allowed for a shuttle launch from Cape Kennedy. Endeavour orbited at a mean altitude of 233 kilometers. Each swath was 225 kilometers wide, with 7-kilometer overlap on each side at the equator. The mission design was such that if the mission performed flawlessly, all landmass between 60°N and 56°S latitudes would be covered at least twice during the 11.2 day mission. Due to convergence of the orbital tracks at the higher latitudes, the swath to swath overlap is greater, and hence the repeat coverage is greater. Figure 3 shows the actual coverage achieved and illustrates the greater coverage with darker shading along the northern and southern limits. It can be seen from Figure 3 that the mission collection was very successful; 99% of the potential area was covered at least once, and 96% was covered at least twice.

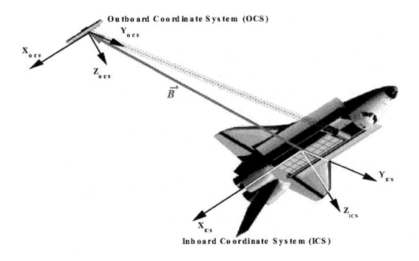

Fig. 2. SRTM in mapping configuration

The radar system consisted of radar electronics and four antennas (a C- and an X-band antenna in the main shuttle bay, and a C- and X-band outboard antenna at the end of the deployable 60 meter mast). The C-Band system used a ScanSAR mode (Figure 4) in which a set of radar bursts were transmitted from the main antenna and then the echoes were received in both antennas. Within each swath there are four subswaths, with HH, VV, VV, and HH polarizations respectively. 'HH' refers to transmittal of horizontally polarized waves and reception of horizontally polarized waves. Similarly VV refers to vertically polarized waves and reception of vertically polarized waves.

Precise interferometry requires knowledge of the absolute position of the shuttle in space and the orientation of the mast. The Attitude and Orbital

Determination Avionics (AODA) system was responsible for collecting the data
needed to compute the attitude and position.

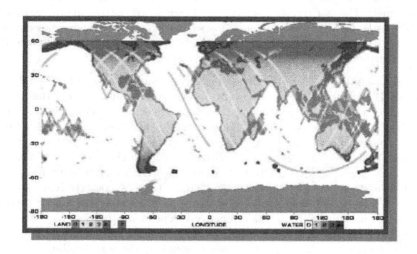

Fig. 3. Coverage of the SRTM Mission

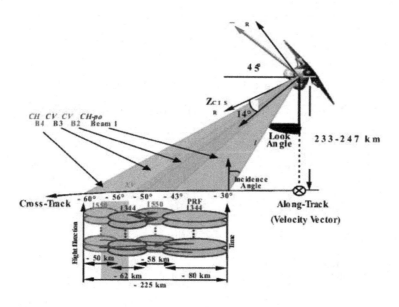

Fig. 4. ScanSAR Geometry

The AODA used an electronic distance meter (EDM) rangefinder to measure the length of the baseline. It used an astro target tracker (ATT) sighted on LED point targets on the outboard antenna to measure the relative baseline yaw angle. The star tracker assembly (STA) measured the overall orientation of the shuttle. GPS receiver antennas located on the outboard antennas measured the shuttle's position in space, with the BLACKJACK receiver boxes located in the Shuttle Bay on the antenna support structure.

Knowing the required relative and absolute accuracy desired for the SRTM based DTED® and the relationships between the parameters and the range equation, the error can be propagated and an error budget determined. Table 1 shows the design error budget allocated among the different sub-systems and error sources. The design absolute accuracy requirements were 16 meters vertical (90% linear error) and 20 meters horizontal (90% circular error). The relative accuracy requirements were 10 meters vertical and 15 meters horizontal. The values in Table 1 are for the far-range sub-swath, the most conservative estimate. The near-range sub-swaths generally have better accuracy because of higher signal to noise ratio. Since the errors are independent, they can be root sum squared (RSS) to yield the total error. Note that the vertical error budget is 22.3m for one pass. Since ascending and descending passes are combined for the final product, this reduces the error by a factor of the square root of 2, bringing the expected error to 16 meters (90%). The actual performance of the shuttle has generally surpassed these specifications.

Table 1. Design Vertical Error Budget

Parameter	90% Error	Induced elevation error in meters
Mechanical Baseline Length	3mm	4.1
Baseline Roll Angle	9 arcsec	15.8
Baseline Yaw Angle	30 arcsec	0.2
Platform Position	1 meter	1
Range	3 meters	1.6
Doppler	30 Hz	0.3
Timing Errors	100 μsec	0
Random Phase	Varies	13
Caltone Phase	8 degrees	7.6
Propagation Errors		0.5
Reference Height	1 meter	1
RSS - Total one pass error		22.3

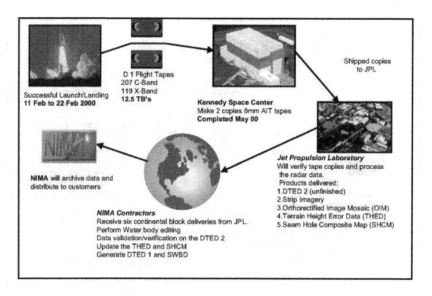

Fig. 5. SRTM Data Processing Flow

4   Data Processing

The SRTM collected 12.5 Terabytes of radar and support data on 326 SONY tapes (207 tapes were C-Band). The general process flow of data handling is illustrated in Figure 5. After the mission, the tapes were copied to Advanced Intelligent Tape (AIT) at Kennedy Spaceflight Center and delivered to JPL for processing. The Ground Data Processing System (GDPS) consists of three parts, the Terrain Processing Subsystem (TPS), the Mosaic Processing Subsystem (MPS), and the Height Quality Subsystem (HQS).

The TPS processes the initial strips of data. These strips initially consist of phase data of the radar returns, which must be "unwrapped" to derive the height data. TPS has to ingest this data and support data to solve for the post heights and correct for any tilts caused by changes in orientation and length of the mast. The MPS ingests the strip data and mosaics them into one product. The original strip points have to be regridded from the original non-uniform distribution to the regular spaced posts of DTED®. Strips are adjusted between one another to produce smooth boundaries. The HQS does the final quality checks for seams, voids and gaps, and spikes and wells. Spikes and wells are not removed in the SRTM DTED® unless evidence exists that the spikes or wells are anomalies. The diagram in Figure 6 illustrates the GDPS system architecture and process flow.

The data delivered by JPL to NIMA is the unfinished DTED®. NIMA's contractors will perform the product finishing. This final processing step includes

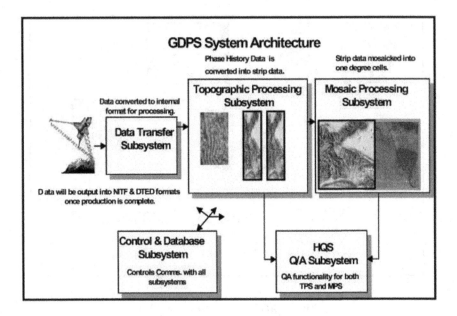

Fig. 6. GDPS System Architecture

adjusting the heights from the ellipsoid to the geoid (World Geodetic System 1984) to make elevations, flattening the water bodies (putting all the water posts at a constant elevation one meter below the lowest shoreline), ensuring drainage runs monotonically downhill, spike and well editing, and general quality assessment.

NIMA manages twin efforts, providing general oversight and performing independent data quality assessment. NIMA evaluates the JPL derived error estimates and assesses SRTM data characteristics.

5   SRTM Deliverables

The SRTM deliverables will include the following: DTED® 1, DTED® 2, the Orthorectified Image Mosaic (OIM), Terrain Height Error Data (THED), Seam Hole Composite Map (SHCM), and SRTM Water Body Data (SWBD). During the post processing period or after DTED® has been finished by NIMA's contractors, a Vertical Obstruction Candidate file for use in generating Vertical Obstruction (VO) data may be generated.

As a result of the SRTM mission a worldwide (within 60°N and 56°S) DTED® 1 dataset will be available to the public. Currently a DTED® 0 set is available (see [5]), which has about a one kilometer spacing. In Figures 7 a&b there is a comparison of a DTED® 0 dataset versus a DTED® 1 set. Immediately apparent is the great increase of data available, approximately 100X the data per unit area.

The main deliverable of the SRTM mission is the DTED® 2 dataset, which at the equator has a nominal 30 meter spacing. The absolute accuracy specifications for the SRTM DTED® are 20 meter CE90 horizontal error and 16 meter LE90 vertical error. DTED® 1 will be created from the DTED® 2 by decimation of the posts. This is done rather than averaging to keep matching posts between DTED® 2 and DTED® 1 at the same elevation. If decimation were not performed, then water body shorelines and flattening derived from DTED® 2 would not match that for DTED® 1. SRTM derived DTED® has different properties from photogrammetrically derived DTED®. Photogrammetric post heights are defined as the measured height of the point at that post. SAR interferometry post heights are derived from the measured average height of the pixel around the post. SRTM DTED® is a measure of the reflective surface, which in the case of C band penetrates into the tree canopy but does not reach the bare earth. Photogrammetrically derived data is often smoothed by the operator during collection as well as in later processing. Its error is generally dominated by a systematic component. SRTM data error will be more dominated by random error. Even though the data can be quite precise, upon close examination a flat surface can exhibit a "popcorn" appearance. This was of concern before the mission, but an adaptive filtering procedure is being developed to mitigate this problem. Overall the SRTM appears to have met or surpassed its error specifications. The error will be further described in the THED description below.

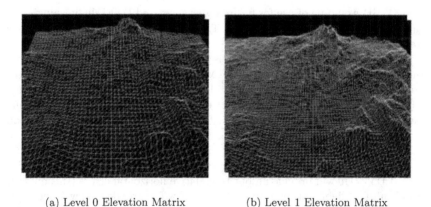

(a) Level 0 Elevation Matrix   (b) Level 1 Elevation Matrix

Fig. 7. Elevation Matrix Density Comparison

The OIM will actually consist of two mosaics, one made from the ascending passes and one made from the descending passes. It will consist of 30 meter pixels, making it approximately the same resolution as a Landsat image but with higher planimetric accuracy. One of the primary immediate uses of the OIM will be to aid in flattening of water bodies in the DTED®. In the SRTM images water generally appears as a void because the radar signal is reflected specularly. Smooth flat areas also can show this effect, as well as radar shadows and areas with layover. A land cover product containing water body information will exist but it is about ten years old. Between the OIM which is current and planimetrically accurate and the water mask derived from the land cover, we believe there is enough information to identify the water bodies among the other similar looking features in the OIM. Once the water bodies over 600 meters minimum width are identified and delineated, corresponding posts are set to a constant elevation. The OIM is nearly global, and because the radar penetrates cloud cover and all but the thickest rains, the tropics are extensively covered. For remote parts of the world these may be the best image maps available.

The SRTM Water Body Data (SWBD) is a vector product that represents the water bodies found within the finished DTED® 2. NIMA contractors will use a combination of the SRTM DTED® 2, the Landcover Water Layer, and the OIMs to accurately generate the water body data. The criteria for delineating the water bodies in the SWBD are:

- Ocean Shorelines;
- Lake/Reservoir boundaries (length greater than or equal to 600 meters and width greater than 183 meters);
- Double Line Drain boundaries (length greater than 600 meters and width greater that 183 meters).

The SWBD would be of use for future radar generated DTED® finishing applications, and as a companion set to SRTM-derived DTED® (quick reference to water portray al).

The Seam Hole Composite Map is a color coded raster map that shows the location of each seam in the SRTM Strip Orthorectified Image data set, holes (voids) in the data set, and the location of voids that have been filled. This product will be updated to depict both post processing and data finishing induced voids. Anomalies in the data for example can be compared to seam locations.

A DTED® cell is typically characterized by four accuracy values: absolute horizontal, absolute vertical, relative horizontal and relative vertical errors. The SRTM DTED® will provide all but the relative horizontal errors (this quantity is difficult to reliably estimate since there are few features defined well enough to do so). The absolute horizontal error will be provided on a continental basis; the estimate will be an output of the processing in the MPS. A significant feature of the SRTM DTED® will be a comparatively detailed characterization of the vertical error. The THED is a file structured similarly to the DTED®, which provides an estimate of the random error at each post. In addition there is a vertical systematic error model (VSEM) derived from comparisons of the derived DTED® post height values with ground truth provided by control points, small patch DEMs provided by NIMA, and GPS kinematic transects. The VSEM will provide estimates of systematic error spectra on a sub-cell basis. The VSEM and THED will allow not only the standard cell vertical accuracies to be estimated, but will also permit customized sub-cell and individual post vertical accuracy estimates to be obtained. Sub-cell absolute and relative vertical accuracy estimates will be included in the VSEM.

Vertical Obstruction (VO) data consists of manmade cultural features such as buildings and power pylons. In Figure 8, from an SRTM image of the Mojave Desert, a string of power pylons is clearly visible across the upper part of the view. Because VO are part of the reflective surface, they are left in the DTED®. It is possible to search the SRTM data for VO candidates that might then be verified by other means. For many parts of the world the SRTM dataset may be the only VO information we have for years to come. Figure 9 is an image of Pasadena California that combines two of the SRTM data types. The image brightness corresponds to the strength of the radar signal reflected from the ground, draped over an SRTM digital elevation model. The scene contains about 2300 meters of total relief. White speckles on the face of some mountains are holes in the data caused by steep terrain. Phenomena called layover and shadowing are responsible for these gaps. In layover, due to the range nature of the radar, tall objects appear to lean towards the view center. While the interferometry process can correct for this, data is lost where the object "layed over". These will be filled using coverage from an intersecting pass.

All the SRTM data is unclassified, but not all is publicly releasable. Available to the public, probably through the United States Geological Survey, will be phase history data, strip imagery and DTED®1 and 2 over the United States. DTED® 1 and strip imagery will be publicly available over the world.

6   Final Remarks

The SRTM mission collected highly specialized data that will allow NIMA to increase the United States Government's coverage of vital and detailed digital terrain information from less than 5% to 80% of the Earth's landmass. This exceeds results from nearly 30 years of efforts with a variety of systems working against persistent cloud cover worldwide. This vital contribution significantly increases the availability to the public of 3 arc-second spaced data worldwide. For the first time in history, a very consistent set of DTED® 1 and 30-meter radar strip imagery all collected from a single sensor in the same 10 day period will be available to everyone. This near-global data between 60°N and 56°S latitudes covers an area where 95% of the Earth's people live. It truly represents " A snapshot of the Earth at the beginning of the 21st Century" which will benefit all of us for many years to come.

Fig. 8. SRTM image of the Mojave Desert

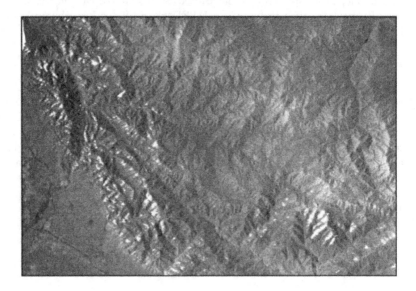

Fig. 9. Shaded Relief with Radar Image Overlay

References

[1] J.C. Curlander and R.N. McDonough, Synthetic Aperture Radar Systems and Signal Processing, John Wiley and Sons, New York, 1991.

[2] F.M. Henderson and A.J. Lewis, Editors, Principles & Applications of Imaging Radar, Manual of Remote Sensing, Third Edition, Volume 2, Published in cooperation with the American Society for Photogrammetry and Remote Sensing, John Wiley and Sons, New York, 1998.

[3] T.H. Dixon, Editor, SAR Interferometry and Surface Change Detection, University of Miami Rosenstiel School of Marine and Atmospheric Science Technical Report TR 95-003, July 1995.

[4] F. Li and R.M. Goldstein, "Studies of multibaseline spaceborne Interferometric Aperture Radars", IEEE Trans. Geosci. Remote Sensing, 28, pp 88-97, 1990.

[5] W.J. Senus, "Global Mapping Shuttle Radar Topography Mission (SRTM)", 4th Global Spatial Data Infrastructure Conference, Cape Town, South Africa, March 2000

[6] T.G. Farr and M. Kobrick, "Mapping the World in 3-D: The Shuttle Radar Topography Mission", Pecoral4 Landsatellite Information III, Denver, Co, December 1999.

Corner, End, and Overlap "Extrusion Junctures": Parameters for Geometric Control

Caroline Y. Westort

The CIM Institute for Applied Computer Science and Industrial Technology (**iCIMSI**)
University for Applied Sciences of Southern Switzerland (**SUPSI**)
6928 Manno, Switzerland
www.cimsi.cim.ch
tel: +41 91 610 8988
westort@cimsi.cim.ch

Abstract. This paper seeks to make explicit the geometric parameters residing at corners, ends and overlap areas of extrusion operations, i.e. "extrusion junctures". Their geometric complexity is discussed in the context of *a generic digital sculpting tool definition* which consists of four components are: (1) blade shape, (2) path shape, (3) blade-path relationship, and (4) Effect. First the blade-path relationship is described in detail, and then the geometric results at these junctures of extruding a blade shape against a path are described, highlighting those factors influencing user geometric control.

1   Introduction

Digital earth moving means digital re-grading of topography. Re-grading means changing topographic form, and geometric control over topographic shape is key to the process. Geometric modelling of terrain has emerged through computer aided design (CAD), which has dealt with the manipulation of libraries of primitive shapes, but large datasets and associated attribute information characteristic of landscape has been the traditional turf of geographic information systems (GIS). The unique problem posed by digital earth moving is that these two traditionally disparate ways of approaching the problem merge. Users want to be able to sculpt a digital 3D surface into any shape, edit and change that shape quickly and easily, and also quantitatively accurately. The modelling technique, *extrusion*, which takes a two-dimensional profile, or 'blade', sweeps it along a three-dimensional trajectory, or 'path', and leaves behind a modelled surface in its wake, is a common technique offered by several commercial 3D software packages [1][2]. Extrusion possesses several important characteristics which enhance a user's geometric control.

Two lines – one for a blade and one for a path – when taken together are an alternative mathematically (i.e. geometrically) equivalent representation for a three-dimensional surface. For every blade-path pair, exactly one surface results, and this is useful as far as geometric control is concerned. This strength is further served by the realization that two lines extruded against each other may produce *any geometry*. The abstraction of extrusion is uniquely suited to the digital environment, for computers are easily programmed to produce a surface out of two lines. Modularly changing the intermediate linear representation of a blade and a path can both mimic known geometry achieved from earth moving and construction equipment, i.e., equipment

C.Y. Westort (Ed.): DEM 2001, LNCS 2181, pp. 78–86, 2001.
© Springer-Verlag Berlin Heidelberg 2001

that works on actual earth, and it can also mimic changes traditionally achieved with representations of landform, like contour lines. That complex three-dimensional geometry can be achieved with such spare and "intuitive" means is powerful. It is far simpler to manipulate two simple lines than the end surface itself. No longer is the three-dimensional geometric change coupled to the desired end output representation. These characteristics of extrusion hold whether the modelling surface is a triangulated irregular network (TIN), contour lines, or a raster grid, etc.. A "translator" between the modular lines and the surface is all that is necessary to achieve the resultant surface geometry – a task algorithmically achievable.

Other characteristics of extrusion which present particular algorithmic challenges, however, are what this paper will refer to as "extrusion junctures"; those regions where the ends of the path, the corners of the path, and the "overlap regions" where the path doubles back on itself create geometric complexity and choice. They present a range of geometric options and are handled differently by individual commercial software packages. This paper tries to make explicit these choices at the extrusion junctures and offers parameterization strategies for how to maximize control over their geometry.

Formal complexity at these junctures rests largely with the set of relationships between blade and path and the effect of the application of the blade along the path. To clarify these relationships we present a generic sculpting tool definition, which provides the context for the work presented here., and builds off of the Ph.D. work of the author [4].

2   A Generic Sculpting Tool Definition

The generic sculpting tool definition consists of four elements:

1. Blade Shape
2. Path
3. Relationship between blade shape and path
4. Effect

For a further discussion of this definition, and an implementation, please refer to [4]. The elements that concern us in this effort are three and four: Definition of the blade shape and path relationship, and the result of extruding them against each other, or *effect*.

2.1 Relationship between Blade Shape and Path

The relationship between the blade shape and path can be described as consisting of an *Orientation* and a *Mode*.

2.1.1 Orientation

Orientation of the blade shape with respect to the path poses possibilities which depend on the two factors: Angle, and Track Point Placement. The track point is the point assigned to the Blade shape which is along which the blade travels when it is exteuded down the length of the path (fig. 2).

Orientation Angle
Orientation angle with respect to a path must conform to the geometric constraint of having only one elevation per x, y coordinate position. If the blade shape were to be revolved around the path's track point, for example, no overhangs or tunnels could result.

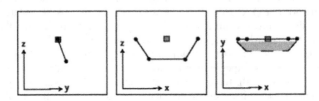

Fig. 1. Orientation Angle. Orientation of the blade shape with respect to the track point.

To illustrate these factors, we consider a path consisting of more than one point. A planar blade shape may experience an orientation angle change with respect to the path as illustrated in figure 1, where in each coordinate plane the angle of the blade shape profile may be moved with respect to the track point.

2.1.2 Track Point Placement
The track point is the point with which the path is aligned as the blade shape is extruded along it. A track point may move, or be placed, anywhere with respect to the blade shape (fig. 2).

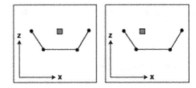

Fig. 2. Track Point Placement. The track point is placed off center in the z-x coordinate plane (a) symmetrical track point placement, (b) track point shifted off-center, i.e., a-symmetrical track point placement

2.1.3 Special Cases
There are two cases of Orientation worth special mention: The first is the *symmetrical blade shape* case, where the shape's plane of symmetry aligns itself with the direction of a path segment. Figure 1 illustrates an example of this case. Notable is how symmetry occurs only in two of the three coordinate planes. The blade shape is not symmetrical with respect to the track point in z-x space, for example. Symmetry greatly simplifies the extrusion functionality

The second special case regarding orientation is when the track point does not lie within the shape's bounding box area. For this case, an *infinite extensibility* may be defined. Indefinite extensibility is an attribute of blade shape. The basic idea is that if a blade shape completely or partially overhangs an area which is far below it, the bottom of the blade shape may extend down to meet the terrain far beneath it. Such an extension may occur in several ways. It may be either a vertical drop, or follow another parameter, like slope. There may also be a parameter like "round the base" for which the edges meet. In traditional contour line use such junctures are referred to as daylight lines, and infinite extensibility would enable one to govern their form. The same applies to blade shapes which are partially or wholly surrounded by existing terrain data which rise far above the blade shape. In this case the blade shape would have to extend up to meet existing grade. This case also applies only for a tracks placed far to the left, or right.

2.2 Static/Dynamic Mode

Another parameter describing the relationship between blade shape and path is static or dynamic mode. Geometric attributes of either the whole blade shape or individual components of it may change *along the course of the path*. Examples of dynamic changes include:

2.2.1 Scale Changes
The blade shape can change absolutely or relative to some criteria.

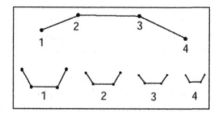

Fig. 3. A blade shape in dynamic mode changing in scale. In this example the blade shape's scale changes along the course of the path, shown in plan at the top of the image, with profiles of the rescaled blade shapes at each path vertex shown in section at the bottom of the image.

2.2.2 Orientation Changes
In addition to scale changes, the Orientation Angle of the blade shape with respect to the path may change. The collection of Orientation variables represents the *state* of the relationship at specified points along the path. The points along the path where the blade shape is placed are intervals. These intervals may or may not correspond with the true path segment points, and may change. Track point may move and Angle of blade shape orientation may also be dynamic along the length of the path.

2.3 Effect

The second component of the *generic sculpting tool definition* relevant to these junctures is *effect*, here defined as the result of applying the blade shape parameter to the path. Exactly how they are applied against each other is key to juncture geometry determination. We discuss the individual cases below.

2.3.1 Corner Extrusion Junctures

For a static blade-path relationship, extrusion of a blade shape along a path consisting of straight, linear the geometric result may be characterized by a cross section of the blade shape oriented perpendicularly to the path along its length. Symmetrical and a-symmetrical track point placement are primary for determining corner extrusion juncture geometry.

Symmetrical Track Point Placement

Figure 4 shows how there are two corner extrusion junctures which need to be set for the symmetrical case. Region *a* is the outer corner, region *b*, the overlap corner between consecutive segments. Region *b* is also discussed in the overlap extrusion juncture section below.

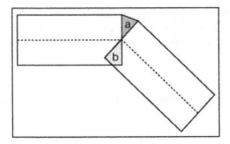

Fig. 4. A simple centrally placed track point corner situation. Two corner extrusion juncture areas to be concerned about; the outer edge *a*, and the overlapping inner edge *b*.

As an example, a blade shape positioned perpendicularly to the track, and in the middle of the z-coordinate plane of the blade shape, at grade with the existing surface, if round corner options are exercised, the corners would result in the geometry illustrated in figure 5. Another example of inner edge *b* being manipulated as an overlap extrusion juncture option is depicted in figure 10.

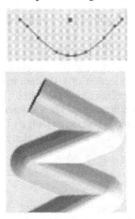

Fig. 5. Example of centrally placed track point with a planar 2-D blade placed perpendicular to the path. Blade and track point are shown at the top of the image. Rounded corner extrusion juncture options are used.

A-Symmetrical Track Point Placement at Corners

The case of an off-center track point placement (figs. 6, 7) complicates the corner situation because instead of only two areas of concern (a & b), there are now at least five for a single corner. And what about the next corner. Since no real-life bulldozer or re-grading tool would create such geometry, the user must now decide on the desired geometry for all of these junctures. For example, should one eliminate the *e* regions? Round the corner as *d* depicts? Or leave it flat with just *a*? These are the parameters which would require handles for a user to govern properly.

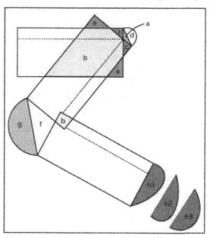

Fig. 6. Diagram of Corner and End extrusion junctures for an a-symmetrical path.

An example of applying this extrusion on an actual surface is illustrated in figure 7.

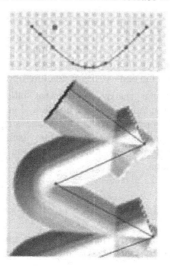

Fig. 7. Example of a corner juncture where the track point position is placed off-center. The track point is found in the blade image at the top.

Other corner shapes

Of course there are many alternatives to rounded corners, including the example in figure 8, where flat corners with a symmetrically placed track point, and planar blade shape are depicted.

Fig. 8. Example of a flat corner extrusion juncture.

2.3.2 End Extrusion Junctures

End extrusion juncture options primarily govern the beginning and close of an extrusion operation. For example in figure 6 above, should a round corner of a blade shape whose track point is placed off center take on shape *h1*, *h1*, or *h3*? If the blade shape is volumetric, the ends may reflect a particular orientation of the volume. Which fraction depends on the end extrusion juncture option desired?

End junctures are of course greatly simplified for static, planar blade shapes, which do not change in scale, orientation, or track position along the course of the path. As previously noted, a cross-profile through the blade shape suffices along the path segments under these conditions and the volume itself is only expressed at path corners and ends. Alternatively, one may wish for the ends to be flat or round or pointed, or some other shape.

2.3.3 Overlap Extrusion Junctures

There are three distinct cases for overlap extrusion junctures: overlap at corners (a and b in fig. 9), overlap over path segments (*i*), and overlaps with previous passes (*j*).

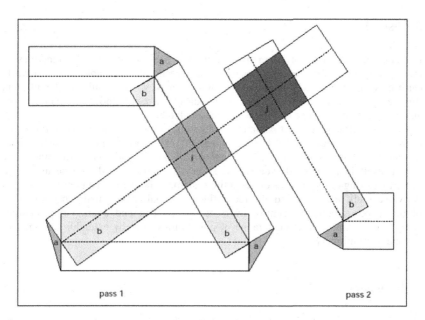

Fig. 9. Overlap extrusion juncture options diagram. *b* represents the overlap option between consecutive segments. *i* represents overlap between non-consecutive segments, and *j* represents overlap between different passes of the extrusion operation.

Overlap junctures where one path overlaps another, either when a path doubles back upon itself, i, or when it's between two separate extrusion passes, j, are easily handled by treating the first segment as existing terrain, and the overlapping segment as the new added or subtracted dimension.

Other Overlap Extrusion Juncture Options
As already briefly mentioned, overlap options govern the geometry between the overlap areas b in figure 1. Example solutions include the following algebraic combinations between existing and new elevation geometry (e.g. +, - , *, min, max, average, >, <). Other solutions may include going to existing grade in these areas, raising all values to a set height, or another function.

Fig. 10. Example algebraic corner overlap extrusion juncture options. Left is an average, Right is a maximum corner overlap option.

3 Discussion

Evident from the range of geometric alternatives possible at extrusion junctures, is that the possibilities in the digital medium far exceed those available with real-world earth-moving equipment. This is an opportunity. The key is to coordinate the different parameters with a system of recognizable handles for a user to most efficiently choose and modify with the range of formal options available.

Given the large number of parameters which may change, there is a need for a kind of metric – or index – for keeping track of, or recording how many of these parameters change at one time. A *complexity index*. In addition to measuring the system itself in terms of degrees of freedom, the metric would also be useful for posterior testing of the success of the entire system. Whether such a metric is feasibly derivable at all, and what kind of direct influence it could have remain open questions of this paper. Geometric control over the corner, end and overlap opportunities enables an extension a broader range of possibilities; thereby extending automation beyond mere mimicry.

References

[1]   Bär, Hansruedi; Interaktive Bearbeitung von Geländeoberflächen; Konzepte, Methoden, Versuche; Ph.D Dissertation, Geographisches Institut, Universität Zürich, (1994)
[2]   Ervin, Stephen M.; Westort, Caroline Y.:   Procedural Terrain; A Virtual Bulldozer; Proceedings CAAD-Futures, International Conference on Computer Aided Architectural Design; Singapore, September, 1995.
[3]   Weibel, R., and heller, M. (1990): "A Framework for Digital Terrain Modeling."; Proceedings of the Fourth International Symposium on Spatial Data Handling, Zürich, Vol. 1, pp. 219-229.
[4]   Westort, Caroline Y.: Methods for Sculpting Digital Topographic Surfaces; Ph.D Dissertation; Department of Geography, Spatial Data Handling Division, University of Zürich, Switzerland, Promovierung: November, 1998.

Contour Lines and DEM: Generation and Extraction

L. Rognant[1], J.G. Planès[1], M. Memier[2], and J.M. Chassery[3]

[1] Alcatel Space Industries, 26 av. J.F. Champollion B.P. 1187
31037 Toulouse Cedex - France
tel: 33 (5) 34 35 69 50
Loic.Rognant@space.alcatel.fr
[2] SINTEGRA - avenue Taillefer
38241 Meylan Cedex - France
[3] Laboratoire des Images et Signaux
LIS/ENSIEG - BP 46
Université Joseph Fourier - Grenoble I
38402 Saint Martin d'hère Cedex - France

Abstract. Contour lines are, as Digital Elevation Models (DEM), 3D information describing the terrain topography. The aim of this work is to demonstrate how the use of a powerful DEM modeling structure, taking into account both representation characteristics, eases their collaboration for topography representation.

After regarding the link between contour lines and DEM, we introduce a new type of triangular representation for surface modeling: *The Delaunay Constrained Triangulation* developed to maintain the Delaunay nature of the final triangulation.

This efficient structure is used for 2.5D DEM design and contour lines management. We present the main properties and interest for DEM modeling with this structure before focusing on the contour lines management.

The bilateral relationship between contour lines and DEM is studied in two Processing chains. First, the use isolines as input data for DEM design. Second, the extraction from triangulated DEM. We propose for both aspects automatic and easy to use processing chains exploiting the Delaunay structure properties.

Keywords: contour line, DEM, DEM design, Delaunay Constrained Triangulation,

1 Introduction

In this paper we study the relation between the contour lines and the DEM to propose an efficient and automatic method for DEM computing from contour lines. This problem has been explored for long time and various solutions have been proposed [5]. Currently contour lines and DEM are hardly cohabiting due to the lack of representing structure taking into account both specificity.

C.Y. Westort (Ed.): DEM 2001, LNCS 2181, pp. 87–97, 2001.

We introduce a the DEM modeling structure : the Delaunay constrained Triangulation very useful for DEM design and management. So it is the natural chose to support processing chain between contour lines and DEM. We briefly present the different types of "Delaunay constrained triangulation" depending on the principles used for its computation. In particular, we introduce stable methods preserving the Delaunay nature of the final mesh.

In a second part we develop the bilateral relationship through two processing chains : a processing chain to compute DEM from contour lines and the dual extracting contour lines from the DEM. We present a simplification pre-processing step of contour lines before developing the DEM design chain preserving the contour lines. Then, we present the contour line extraction using the Delaunay/Voronoi properties of the triangulated mesh.

Finally, we present some applications in numerical geography of DEM and contour lines combined.

2   Contour Lines and DEMs

Contour lines and DEM are both a representation of the same real topography. They just correspond to different ground surface sampling strategies as topographic profiles or grid sampling. So, it should be easy to build a relation between them.

Moreover, contour lines are now considered as an end product. So, they are mainly used for map edition and not for digital geography. In that case, the lines are often "corrected" by the land surveyor to make them easy understandable for the map user.

The current digitalization of map library leads to DEM production from contour lines which are coming from digitalized maps or geographical vector library. This induces the following questions :

- *What kind of link exists between contour lines and DEM?*
- *What kind of topographic representation structure can be used to build and highlight that bilateral relation?*
- *How taking into account this planimetric and height information as it is?*

Currently we use a grid structure to interpolate the contour lines. This produce the usual DEM artifacts due to this computation method :

- Effects of the sampling rate grid : the sampling rate is global and can not take into account the local variations of lines shapes, density, ... this leads for instants to miss the summits and build flat mountain top.
- the interpolation effects on slope, aspect, drainage network.

So it is hard to determine a priori the final DEM quality although knowing the contour line precision. Moreover it require a lot of computation.

On the other and, we can use a triangular representation taking into account punctual data with linear features. This leads to choose the Delaunay constrained triangulation.

3   The Delaunay Constrained Triangulation

3.1   The Problem

The Delaunay triangulation is a well known method for triangular meshing a set of points [6], [9], [2]. The problem of the constrained triangulation is to make appear a constraint graph described by constraint edges. Each constraint edge is then a part of triangles.

Definition 1 (Delaunay stable incorporating method). *A constraint incorporating method in a Delaunay triangulation is said to be stable if the resulting triangulation is still a Delaunay triangulation.*

The difference between the stable and unstable methods can be translated in their naming manner. It has to be stressed that the *Delaunay Constrained Triangulation (DCT)* is different from the *Constrained Delaunay Triangulation (CDT)*. The CDT are produced by Delaunay unstable methods. On the contrary, DCT are the result of a Delaunay stable algorithm.

Those nuances are highlighted when we look at the different constraint methods.

3.2   Delaunay Triangulation under Constraint [7]

The principle is to redefine the building criterion of the Delaunay triangulation. So, we define the "constrained empty circle criterion" taking into account the graph of visibility of the configuration. The final triangulation is no more of Delaunay nature.

3.3   The Unstable Methods

First we compute the Delaunay triangulation of the vertices and the constraint extremities. Then we incorporate the missing constraint edges by re-triangulating the constraint edge's neighborhood while preserving the edge integrity.

Theorem 1 (triangulation without internal points [8]). *For each area whose boundary is a simple non crossed polygonal lines, there exists a triangulation without internal points.*

So, the resulting triangulation verifies exactly the constraint field but is no more of Delaunay type.

3.4   The Stable Methods [12]

The Delaunay stable methods are based on the breaking method building new Delaunay compliant edges.

Theorem 2 ([13]). *In a Delaunay triangulation, for each non-Delaunay compliant constraint edge, it always exists a partition leading to Delaunay compliant edges.*

The theorem 2 leads to different edge partition algorithms splitting the constraint edge to make it Delaunay compliant. In [12], we present some methods more or less expensive because of the new points introduced during the partition computation. The final triangulation is still of Delaunay nature and possessing all its interesting properties.

3.5   Conclusion

The constrained Delaunay triangulation has different methods depending on the hypothesis used to compute it. We can highlight three main classes :

- Delaunay triangulation under constraint (new type of triangulation),
- Constrained Delaunay triangulation CDT (no more Delaunay),
- The Delaunay Constrained Triangulation DCT (of Delaunay nature).

We use the DCT mesh structure for DEM design because of the Delaunay preserved properties of the final mesh.

4   The Delaunay Constrained Triangulated DEM Representation [11]

The triangulated DEM is an alternative to the classical regular DEM. Instead of applying a structure to the terrain (the regular sampling), we allow the terrain to structure the DEM. So, we adapt the data sampling to the type of landscape it represents and to the accuracy needed for the DEM application.

The triangulated DEM includes the properties of standard DEM plus the previous considerations. To keep the terrain structure, we warrant the presence of topographic lines in the mesh by using them as mesh constraints.

Data Used in the Constrained Triangulation Representation

The topographical data are considered to be a set of 3D scattered data including a set of topographical constraint lines (watershed, talweg).

This allows to use a small amount of points and an irregular (spatial) distribution while preserving the DEM realism and quality. Moreover, the model can be refined for the most important area whereas regions of lower interest are more roughly described.

DEM Design

The DEM is considered as an open surface with neither cavity nor pleat. This allows us to use a 2D Delaunay Triangulation method to reconstruct a 2.5D surface. All the DEM constraints are incorporated as mesh elements.

Some Useful Properties for DEM Design

Multi-sources : The data are from various origins : radar images, optical images, maps.

Irregularity : The irregular structure of the mesh and the iterative insertion mode allows to handle data at different sampling rates. For instance, this structure can include a TIN mesh and a regular grid mesh.

Iterative algorithm : this algorithm allows the dynamic update of the DEM, inserting or deleting vertices. This property is also used for multi-scale representation and for moving or zooming (to have a local higher scale) a region of interest.

Constraints : The use of constraint lines appearing as edges in the DEM mesh is a guarantee of good realism for a low number of necessary points. This helps modeling a wide area at low cost.

5   DEM and Contour Lines

We consider starting from digitalized contour lines as for example from the altimetric data base of the French national geography institute (IGN) or a Vector Map (VMAP from the National Imagery and Mapping Agency (NIMA)) contour lines layer. We don't speak here about the digitalization of contour lines from a map.

Hypothesis (2.5D contour lines). A contour line do not cross other lines from different altitudes.

5.1   Contour Lines Pre-processing

We must note that often the contour lines are corrected during the map edition to make them easy understandable for the customers. The pre-processing step can be seen as this data preparation.

The pre-processing step allows lowering the data amount while adjusting their quality and precision to the final product specification. Because the contour line simplification is not a sub-sampling process, it is treated as a data compression (with lost) problem. So, the processing is divided in three steps :

- data "characterization" by the line complexity analysis,
- simplification methods,
- quality control during the simplification.

These methods are applied in a first time to single contour line, then we apply it to a set of lines.

Contour line complexity: The aim of contour line complexity calculation is to help choosing either the simplification method or the simplification parameters.

Definition 2 (Contour line complexity). *The contour line complexity is a measurement describing the trajectory variability of the line.*

We developed various complexity measurements. We present here the "sinus" complexity.

$$Cplx = \frac{1}{\text{number of points}} \sum_{i=0}^{nbPts-1} \sin^2(\theta_{i,i+1}) \quad Cplx \in [0,1]$$

where θ_{ij} is the angle of direction change between the edge i and the edge j. Because the complexity is evaluated before and during the simplification, we assume that θ_{ij} is small due to the fine over-sampling of the line. A null complexity corresponds to a straight line. The complexity increase with the number and the importance of direction change in the line.

Fig. 1. a "simple" line $(Cplx = 0,035)$ / a "complex" line $(Cplx = 0,07)$

Simplification methods: We developed three class of simplification methods:

"In the water course". The line is simplified in one pass from start to end. It is a very quick method but less efficient than the others.

Iterative methods. The method repeat the treatment over the complete line until the quality criterion is obtained. It is an efficient method analyzing all the arcs composing the line but it is slow.

Recessive methods. The contour line is split in two sub-arcs which are separately simplified. It is the most efficient method adjusting the simplification to the local complexity.

The figure 2 illustrate the different orders of simplification of a single line with a recursive method. We note a first "purifying" step going from 350 to 56 points (84% gain) where the line is simplified without significant quality loss. This corresponds to a high frequency noise reduction. After this, the quality loss increases. The first purifying step is especially interesting for hand digitalized lines.

The simplification method can be refined by the use of a combination of different methods for the same isoline.

Fig. 2. Different order of simplification of a single contour line

Simplification quality: When the lines are simplified and during the simplification process, we need to estimate the quality loss. A too much simplified
contour line may have intersections with its neighbor lines (what is excluded by
hypothesis 5) whereas the line can be more simplified to comply with the DEM
specification. A too important simplification leads to artifacts apparition due to
the loose of the original lien shape.

We studied quality criteria that must reflect the relation between the original
line and the simplified line. So, we use for instance the area appeared between
the lines (fig. **??**), the length loose, the compression rate , ... :

$$q = \frac{\text{lost area}}{\text{initial length}}$$

The more q is low, the more the lost area between the lines is low. By experience,
a good simplification quality is under 2.

Multi-line simplification: The problem of the simplification a global set of
contour lines is different from single line simplification. This corresponds in a way
to topography simplification. Often singles simplifications, even of good quality,
may produce intersections between lines by the line displacement in the neighborhood of the original line. This depends especially of the line density, which
can be directly related to the terrain topography. The multi-lines intersection
problem is treated in two ways :

- preventing the lines intersections by choosing simplification methods, strategies and parameters,
- a post-processing step by correcting the introduced artifacts.

To prevent the lines intersections, we can for instance :

- use of master contour lines less simplified and staying untouched by the
  correcting process.
- use of the contour lines density: For instance, in a plain the contour line are
  widely separated and simple so they can be more simplified while in mountain
  area they are very dense and complex and should be less simplified.
- use of simplification direction from top to down heights.

In a second time, we must correct the defaults appeared between the contour
lines. We developed different methods, which are not equivalents due to the
different leading principles :

- interpolation of new points for the lowest contour line avoiding the intersection. This method may propagate the modification the all the next lower lines.
- local point exchange between the lines.
- the local un-simplification of the lines using the 2,5D assumption saying that the "original" lines do not have intersections. This is the safer and accurate correcting method but may re-introduce lots of points.

Remark 1. "Purified" contour lines introduce no intersection problem due to the high quality of the single simplification.

5.2   Contour Lines DEMS

We present a processing chain to produce automatically DEM from contour lines. In a first step, the contour lines are pre-processed to respond to the DEM specifications : accuracy, amount of data. The pre-processing is conducted in two ways :

- a selection of contour lines to obtain the height accuracy related to the altitude sampling rate of the lines,
- a contour simplification to adjust the data amount and the planimetric accuracy.

Then the contour lines are treated as linear constraint for the triangulation computation. This data needs to be completed because of the height sampling rate leading for instance to plateau on mountain's top while it exist a punctual summit. So the data may be completed by punctual information (as summit) and linear networks (as talweg, slopes) (fig. 3).

This data management is allowed by the Delaunay Constrained triangulation chosen for DEM design. The 2.5D nature of the reconstructed surface induce that the DEM can not describe overhang.

Fig. 3. DEM design from contour lines (left: the data = contour lines + talweg + summit; right the final DEM)

The DEM is automatically computed directly from the contour lines data. The use of contour line as constraints avoid misplaced triangles connecting two lines separated by an other. Finally, a post process detect the plan triangles. This step is man supervised in order to detect the real errors. Two correcting methods are used :

- a manual method where the operator add a slope profile,
- an automatic method computing the steepest slope constraint by interpolation between the two lines [1].

The obtained DEM is a triangular structure where the contour lines appear as mesh elements. Because it still is a Delaunay triangulation, this DEM inherits of all the Delaunay properties and can by example be easily resampled (at any sample rate) into a regular grid DEM.

5.3   Contour Line Extraction

Contour lines extraction from TIN are well known [15]. They use general triangulated mesh without taking into account its properties leading to extra computation as triangle ordering. In our case, We consider that the DEM is a Delaunay triangular structure. We use the Delaunay dual diagram, the Voronoi diagram, to help the contour line extraction. The Voronoi cells give the neighborhood of triangles accelerating the extraction process, by introducing a snake behaviors to the contour line extraction.

Remark 2. If the DEM was computed from some master contour lines, they must be found by extraction exactly as they were during the previous step.

The "raw" contour line obtained can be post-processed to enhance their look and shape for map edition for instance. The lines can be smoothed by spline interpolation using different spline functions: B-spline, weighted spline. This step requires being careful to avoid intersections dues to the spline behavior.

5.4   Applications of DEM and Contour Lines Relationship

The processing chains we presented lead to lots of applications using the duality between contour lines and DEM. We can cite :

the re-interpolation of contour line : The interpolation of new contour lines from an original set of isoline needs to build a relation between the line. This lead in a certain way to compute the contour line DEM for the relation building and the extraction process corresponding to the isoline interpolation. The interest of Delaunay constrained triangulation representation is the best approximation property [10] leading to the best mesh for interpolation. Moreover, we can compute a continuous, derivable DEM surface from such triangulated DEM allowing using a snake behavior for the isoline interpolation.

the fusion of contour lines and other sources : mixing contour line over a huge surface with precise or high resolution DEM (regular or not) for the study of watersheds basins.

the topography simplification : The topography representation by contour line is already a sort of surface simplification. First, we build the DEM from all available data (grid DEM, contour line, GPS points, network, ...).

Then, by contour line extraction the topography is simplified preserving the local precision of the computed DEM. As said in [1], [14] the contour lines implicitly contains the ridges and drainage lines which are essential for the terrain surface description. This line set can be completed by VIP (Very Important Points) [16] describing the punctual part of the terrain surface description. Later, a new DEM can be computed from the extracted data.

quick DEM computation for an iterative process : Lots of remote sensing or geographical processing are improved with a topography a priori (furnished by a DEM). We can quickly propose a simple and low cost DEM computed from contour lines. The conservation of the Delaunay nature allows DEM refinements with other data.

DEMs co-registration by contour line extraction : Two DEMs with different characteristics (scale, sample rate, ...) can be easily co-registered by extraction a the same rate of contour line sets that will be superposed.

6   Conclusion

We can say that contour lines are no more an end product and they can be easily used in a bilateral relation with DEM for design and extraction. This opens new opportunities to use them in numerical geography. Through the Delaunay Constrained Triangulation (a particular surface and DEM representation taking into account contour lines and DEM characteristics), We have presented three processing chains :

- a pre-processing chain to simplify contour lines before their use. This allows managing the requested quality (for DEM design, or topography simplification) toward the original one.
- an automatic and easy to use processing chain for DEM design. The contour lines are preserved in the final product; so they give their quality and precision to the computed DEM.
- an automatic and easy to use contour line extraction process, computing the lines from a Delaunay triangulated DEM.

This opens new opportunities of applications using the duality between DEM and contour lines.

References

1. Aumann G., Ebner H. and Tang L., "Automated derivation of skeleton lines from digitized contours", International Archives of Phototgrammetry and Remote Sensing, 1990, 28(4), pp. 330-337.
2. Bertin E., "Diagrammes de Voronoi 2D et 3D : application en analyse d'images", Ph.D thesis, TIMC - IMAG, Université Joseph Fourier - Grenoble 1, 1994.
3. Brandli M., Schneider B., "Shape modelling and analysis of terrain", International Journal of Shape Modelling, vol. 1, No. 2, 1994, pp. 167-189.

4. Chew L. P., "Constrained Delaunay Triangulation", Algorithmica, vol. 4, no. 1, pp. 97-108, 1989.

5. Christensen A.H.J, "Fitting a triangulation to contour lines", in Auto-Carto 8, 1988, Baltimore, pp. 57-67.

6. Edelsbrunner H., "Algorithms in combinatorial geometry" Springer Verlag, 1988.

7. De Floriani L., Falcidieno B. and Pienovi, C., "Delaunay-based representation of surfaces defined over arbitrarily shaped domains", Computer vision, graphics and image processing, vol. 32, pp. 127-140, 1985.

8. George P. L. and Borouchaki H., "Triangulation de Delaunay et maillage - application aux éléments finis" Hermès (eds.), 1997.

9. Preparata F.P. and Shamos M.I., "Computational geometry - An introduction" Springer Verlag, 1985.

10. Rippa S., "Minimal roughness of the Delaunay triangulation", Computer Geometric Design, vol. 7, pp. 489-497, 1990.

11. Rognant L., Chassery J.M, Goze S., Planès J.G, "Triangulated Digital Elevation Model: definition of a new representation", Proceedings of ISPRS'98, Stuttgart, 2-8 September.

12. Rognant L., Chassery J.M, Goze S., Planès J.G, "The Delaunay constrained triangulation: the Delaunay stable algorithms", Proceedings of CAGD'99, London, 14-16 July.

13. Rognant L., "Triangulation Contrainte de Delaunay: application a la représentation de MNT et a la fusion de MNT radar", Ph.D thesis, Laboratoire des images et signaux (LIS)/Institut National Polytechnique de Grenoble (INPG), Université Joseph Fourier - Grenoble 1, 2000.

14. Schneider B., "Geomorphologically sound reconstruction of digital terrain surfaces from contours" Proceedings of the 8th Symposium on spatial Data Handling, Vancouver-Canada, 1998, pp. 657-667

15. Van Kreveld M., "Efficient Methods for Isoline Extraction From a TIN", Int. J. Geographical Information Systems, Vol 10, No. 5, pp 523-540, 1996.

16. Chen Z. and Guevara J.A., "Systematic selection of very important points (VIP) from digital terrain model for constructing triangular irregular networks", Proceedings of Auto carto 8, Baltimore, 1987, pp. 50-56

17. Ware M.J., "A procedure for automatically correcting invalid flat triangles occurring in triangulated contour data", Computers and Geosciences, 1998, 24(2), PP. 141-150.

Modeling of Ecosystems as a Data Source for Real-Time Terrain Rendering

Johan Hammes

PO Box 1354
Stellenbosch
7599
South Africa
jhammes@mweb.co.za
Fax: +27 21 880 1936
Telephone: +27 21 880 1880

Abstract. With the advances in rendering hardware, it is possible to render very complex scenes in real-time. In general, computers do not have enough memory to store all the necessary information for sufficiently large areas. This paper discusses a way in which well-known techniques for modeling ecosystems can be applied to generate the placement of plants on a terrain automatically at run-time. Care was taken to pick algorithms that would be sufficiently fast to allow real-time computation, but also varied enough to allow for natural looking placement of plants and ecosystems while remaining deterministic. The techniques are discussed within a specific rendering framework, but can easily be adapted to other rendering engines.

Keywords: ecosystem modeling, ecotope modeling, compression, real-time rendering, terrain modeling, terrain visualization

1 Introduction

With the advances in rendering hardware, it is possible to render very complex scenes in real-time. Scenes of up to 200 000 triangles are becoming possible with much more complex scenes looming in the near future. The ability to render very complex scenes necessitates databases that can supply the rendering engine with enough data.

Landscapes for flight simulators, games or visualization, poses a problem due to the large amounts of plants found on them. It is not uncommon for a flight simulator to have 500 million big trees in its database area, and countless smaller trees and shrubs. Due to the scale, and the amount of memory needed to hold the information, these trees have in general been left out of the simulation, or incorporated into the textures. In this paper, a system of ecotope[1] modeling, that calculates the plants in the view frustum in real-time, is presented.

The exact placements of these plants are seldom known, and seldom of real importance. For vast areas of countryside they add to the general feeling of realism.

---

[1] Ecotope: A particular habitat within a region with relative uniform climatological and soil conditions. Typically, specific ecotopes will be associated with specific ecosystems.

C.Y. Westort (Ed.): DEM 2001, LNCS 2181, pp. 98-111, 2001.
© Springer-Verlag Berlin Heidelberg 2001

This lends itself to the modeling of ecosystems to determine the placement of plants. By modeling the ecosystems, and generating only the plants in the immediate vicinity, the overhead of storing millions of plants in memory can be avoided.

Ecotopes are a good predictor of ecosystems. Rather than modeling the ecosystems (modeling the complex interactions between plants), ecotopes are modeled and the landscape is populated with representative ecosystems.

Fig. 1. Two images created by Sam Bowling, using the World Construction Set [1] by 3D Nature. It shows the realistic natural scenery that can be generated with ecotope modeling

By using the basic parameters of elevation, relative elevation, slope, slope direction and multi fractal noise, it is possible to generate ecotope information that can be used to predict the ecosystems. An ecosystem is assigned a probability that it will exist as a function of the above five parameters. By evaluating them at a particular position in the world it is possible to find a probability for each ecosystem. The ecosystem with the highest probability is assigned to that particular area.

The appearance of each ecosystem is defined in advance. This includes descriptions of the type, frequency and placement of plants, as well as typical ground cover. This information is used to build a representation of the ecosystem. The information is passed on to the rendering engine where it is cached, and reused for all subsequent frames that look at the same area.

In section 3, the framework that is used to render the scenery is discussed. The definition of ecosystems and modeling of ecotopes is structured around this. Section 4 looks at common parameters that is used to model ecotopes and ecosystems. In section 5 this is extended to show how ecosystems can be constructed to fit into the rendering framework. In section 6 the algorithms that is used to calculate the ecotopes and build a representative ecosystem is presented. I also look at some optimizations that are necessary for real-time performance. Section 7 looks at the results and is followed by a conclusion in section 8.

2 Background

Most of my previous work dealt with large-scale terrain visualization for both commercial, gaming and military flight simulators. As rendering speeds increased it became possible to add more and more objects into the terrain adding to the realism. At first it was adequate to develop off-line tools to place objects and store them in the database. With the current commercial rendering capacity far exceeding one million triangles per second, it is impossible to store all the objects in a limited amount of RAM. A way had to be found to efficiently compress this data and extract only the data in the immediate vicinity of the camera in real-time.

When viewing natural features it is the general patterns that define the area, rather than the exact detail. The specific position of a plant does not define a natural scene, but the relative placement due to competition for natural resources is very important. Ecosystem and ecotope modeling constitutes a form of compression for natural environments. While the exact position and type of plants are not preserved, the general statistical properties remain, allowing a representation to be built with the same feeling and character.

A number of commercial programs that model ecosystems (ecotopes) to synthesize data exist. World Construction Set (WCS) by 3D Nature [1], Terragen by Planetside Software [2], and Genesis II by Geomantics [3] to name a few. WCS is one of the best examples of ecotope modeling as seen in Figure 1. The realism in the images is a convincing argument that ecotope modeling is a good way of generating natural scenery.

While these programs can render very realistic scenery, they are not real-time, usually taking minutes to render a picture. This paper looks at the possibilities to adapt ecotope modeling to the time constraints of real-time rendering.

3 Rendering Framework

The rendering framework divides the area into a set of square tiles, that can recursively break into smaller tiles in a quad-tree structure. Tiles split at a fixed distance from the camera, relative to their actual size and the field of view of the camera. This result in tiles that are roughly the same size in view space. Figure 2 shows the way that the surface splits for a particular camera position close to the ground. This splitting can be stopped at any level, and is combined with a prediction algorithm that looks at the movement of the camera to predict tiles that will be visible in the near future. There are a lot of tiles present that do not fall within the bounds of the view frustum. This is due to the cache prediction algorithm. A fast rotation of the camera is the most difficult situation to handle, and requires a lot of tiles adjacent to tiles in the view frustum to be present.

Only the tiles included in Figure 2 need to be extracted from the database, and reside uncompressed in memory. The algorithm makes use of frame coherence to update only those tiles that change between frames.

Each tile in this representation consists of:

- A 17x17 grid of elevations covering the extents of the tile. As the tiles split and become smaller, the actual resolution of the grid increases. This is the first step (block based) in triangle optimization.
- A 128x128 pixel texture depicting the ground cover of the area. The size was chosen to ensure roughly a 1:1 pixel to texel ratio for screen resolutions between 800x600 and 1024x768. At higher resolutions it will be necessary to either use a bigger texture or change the tile split metric to maintain this ratio.
- A list of objects that reside on this tile (trees, rocks, houses etc). This is only a key showing the type of object and does not include any information on the geometry of the object or how to render it.

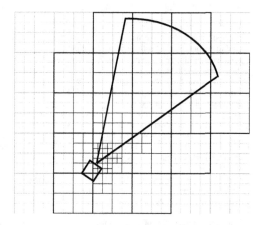

Fig. 2. The subdivision of tiles in the view frustum for a particular camera position

Because the parent tiles always exist, it is possible to render a representation of the terrain even if all of the tiles are not cached. While all four daughter tiles have to exist before any one of them can be drawn, it is possible to draw the parent tile for a few more frames while caching the daughter tiles. This enables the system to update only a few tiles every frame, while maintaining a consistent view of the terrain. It is very important since it allows a fixed time to be allocated to extract a tile from the database. Although drawing a parent tile affects the visual quality (lower resolution) it is possible to handle 400 degree per second turns with only three tiles being updated every frame. This is fast enough for the most demanding games or simulators. Due to the nature of the system a tile will never be cached unless its direct parent tile is already cached. This allows the use of information from the parent when caching a tile.

4 Modeling Ecotopes

This section looks at common variables that are used to model ecotopes in existing software and discusses their influence on ecosystems, and the way in which these parameters are combined to predict ecosystems.

Due to the constraints of real-time modeling it is important to use a fairly simplistic system that can be optimized. Table 1 shows a list of variables being used by WCS [1], with both Terragen [2] and Genesis [3] using a subset.

Each ecosystem in the database is assigned a probability for each of the above variables. By combining the probabilities, a probability for each ecosystem can be determined at a specific position on the terrain. The ecosystem with the highest priority gets assigned to that position.

Table 1. Summary of the variables that is used to calculate ecosystem placement

Elevation	The height above sea level. With increases in elevation, the general conditions become harsher. All plants have an upper limit at which they can survive. Plants also tend to become smaller with increases in altitude.
Relative elevation	Relative elevation refers to the local changes in height, with negative values showing depressions, valleys etc, and positive values showing ridges. This is the higher frequencies of the terrain. Relative altitude affects plant growth since valleys are generally wetter, as well as more sheltered. Ridges on the other hand tend to be exposed to the elements much more.
Slope	The slope of the terrain has a direct bearing on the quality and depth of the soil, as well as water retention due to runoff. Steep slopes tend to have small shrubs and grass cover. Very steep slopes tend to be exposed rock with no vegetation.
Slope direction	The direction that the slope faces has a direct bearing on how many sunlight hours it receives each day, as well as being more sheltered or exposed to the prevailing winds.
Multi-fractal noise	Some plants and ecosystems also exhibit local grouping behavior independent of the above 4 variables. One reason is reproductive behavior. Plants that either drop their seeds, or reproduce vegetatively from roots tend to exhibit strong grouping behavior. A lot of multi-fractal noise functions exhibit similar patterns, and can be used to change the probability of ecosystems, or the density distribution of plants within ecosystems, to model this behavior.

5 Defining an Ecosystem

Unlike most other systems (WCS [1], Terragen [2] etc.) that do a single ecosystem calculation for each position on the terrain, ecosystems are divided into layers to fit into the quad-tree rendering structure, and each layer is solved recursively as the tiles split. This section looks at the way that ecosystems split into layers, the sort of information available on each layer and the ways in which they can be combined to form complex landscapes.

A layer consists of a vegetation canopy and ground cover and is defined for a specific layer in the quad-tree terrain. Since a tile will always fall within a pre-determined size range in screen space, it is possible to estimate the pixel size of all plants on the tile. The vegetation canopy is defined as all the objects (plants, rocks etc.) that are roughly one pixel in size when this layer is first used. Table 2 shows a typical representation for savanna. There is no need to solve for plants that will be smaller than one pixel on screen, since they will make no contribution to the visual quality of the scene, and can be incorporated into the ground cover. Plants that are bigger are incorporated into a higher level and will already be solved at this stage. The ground cover is a texture with a representative image of the ground as well as all plants that are still smaller than one pixel in this layer, and can be seen in Figure 3.

Table 2. Summary of possible layers present in a savanna ecosystem

Layer	Vegetation canopy	Ground cover
0	None	Typical aerial photographs of the terrain with trees shrubs and grass. See Figure 3
1	Big trees	Smaller trees, shrubs and grass
2	Medium trees	Small trees, shrubs and grass
3	Small trees	Shrubs and grass
4	Big shrubs	Small shrubs and grass
5	Shrubs	Grass
6	Small shrubs and tall grass	Grass with patches of brown ground
7	Short grass	Small plants, rock and ground

5.1 An Ecosystem Layer

Each layer in the ecosystem consists of the following items:

- A texture representing typical ground covers. This is used to build the texture that is draped over the terrain when rendering. A number of examples are shown in Figure 3.
- A list of possible objects in this ecosystem. This includes :

    - The type of object. (It is possible to define objects like rocks as well, as long as they have a natural distribution)

    - Object density. This is used in conjunction with a random offset to determine the number of objects in a specific area.

    - Size and color variation information. This is used to generate variations in the appearance of objects, and is particularly useful when billboards are used to render trees.
- A list of all possible layers on the next level that can follow this one (see Figure 4).

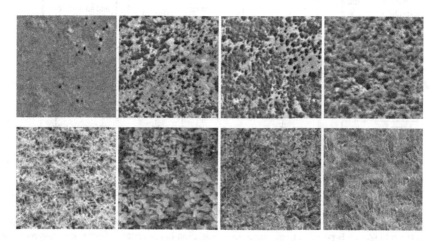

Fig. 3. The top four textures define four possible types of ground cover for savanna, all with different amounts of tree cover. These are layer zero ground covers (Table 2) with no 3D plants present, and were obtained from aerial photographs. The bottom four textures define ground cover on layer eight showing small plants, grass and soil

5.2 Combining Layers into Ecosystems

To facilitate in a diverse environment with the minimum of data, ecosystems are constructed from a tree of possible eco-layers. Each layer in the ecosystem defines which layers on the next level can exist under it. A schematic presentation of this can be seen in Figure 4. While there can be no trees at all in the grassland, it is possible for areas with partial or dense tree cover to have grass underneath them. In general, layers with many plants will have sparse layers under then, while layers with few plants will have more dense layers underneath them due to competition for sunlight.

When an area is rendered, the objects rendered are the sum of all the all the objects on this layer, and all the layers above the current one. No object intersection detection is done. The ground cover used, is the ground cover for the current layer.

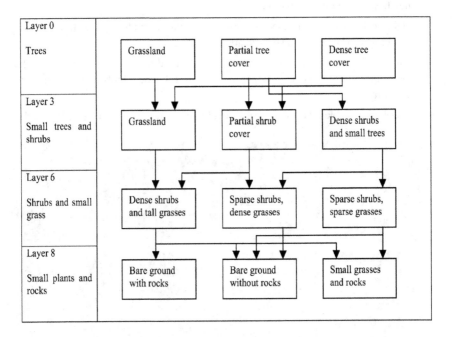

Fig. 4. A schematic presentation of possible combinations of ecosystem layers. By combining different layers it is possible to get much more variation in ecosystems. Some of the layers have been left out for clarity

6 Run-Time Calculation of Ecosystems

The modeling of ecosystems can be divided into the five processes shown in Table 3. This section looks at their specific implementation.

Table 3. Pseudo-code showing the basic operations used to build a tile

```
CacheTile()
{
        BuildElevationGrid();
        // Extract a 17x17 elevation grid from the database
        BuildRelativeElevationGrid();
        // Builds a 17x17 relative elevation grid
        BuildEcosystemGrid();
        // Builds a 17x17 grid of assigned ecosystems
        CalculatePlantLayout();
        // Solve the plants and add them to the tile
        // Section 6.2
        BuildTexture();
        // Build a representative ground-cover texture
        // Section 6.3
}
```

6.1 Building the Elevation Grid

The elevation grid is a direct copy of the elevation data stored in the database (Figure 5a). The database has to define an elevation grid for all tiles on level 0. Most higher level tiles will not have elevation data in the database due to the availability of data and the amount of RAM available to store elevation data.

If no elevation data exist in the database, the elevation grid will be constructed from the parent tile's elevation grid. 9x9 elevations from the parent's elevation grid are copied to every second position in the child's elevation grid. It is interpolated linearly and a small random offset is added to ensure variation. This is similar to a fractal height field generator.

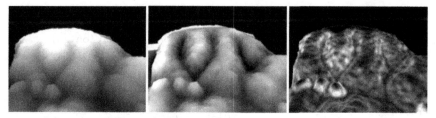

Fig. 5. Screenshots showing (from left to right) elevation, relative elevation and slope as calculated for a mountainous area

6.2 Building the Relative Elevation Grid

Each tile in the database has four average elevation values, one for each corner. These four values are retrieved from the database, and interpolated to generate a 17x17 grid of average elevations for this tile. While the database stores enough information to calculate quadratic interpolation, linear interpolation was implemented for speed. As can be seen in Figure 5b, the results of linear interpolation is convincing. The relative elevation (Figure 5b) is the difference between the elevation and the average elevation.

If a tile does not exist in the database (and its elevation grid is interpolated from its direct parent), no average elevation can be calculated. The relative elevation is calculated by linearly interpolating the relative elevation of the parent tile. This is halved, and perturbed again with the same random offsets that was added when interpolating the elevation grid. While this is not an accurate mathematical solution, it does yield acceptable results while being fast.

6.3 Building the Ecosystem Grid

The ecosystem grid is a 17x17 grid of ecosystem types coinciding with the 17x17 elevation grid defined for the tile. A complete evaluation of the parameters is done for each of the positions in the grid, and an ecosystem assigned to that element. Using a 17x17 grid instead of 16x16 allows for overlap between tiles. By accepting the penalty of re-computing data, tiles can be isolated from each other, and solved with no knowledge of any other tiles in the area other than its direct parent.

Table 4. Pseudo-code showing the calculation of the ecosystem grid

```
BuildEcosystemGrid()
{
        // step through the grid and calculate the
        // ecosystem with the highest probability at
        // each position

        for (y=0; y<17; y++)
                for (x=0; x<17; x++)
                        Eco[y][x] = CalculateEcosystem();
}

CalculateEcosystem()
{
        CalculateSlope();                          // Equation (3)
        for (i=0; i<NumEcosystems; i++)
        {
                CalculateSlopeSkew();           // Equation (7)
                prob[I] = CalcEcoWeight(      // Equations(8,9)
                        elevation + Skew,
                        relative_elevation,
                        slope )
                        + (random offset);
        }
        return (ecosystem with the highest probability);
}
```

Calculate Slope. A 17x17 grid of slope values in both the x and y directions are defined by equations (1) and (2). The slope is defined in equation (3) and the result can be seen in Figure 5c. Unlike the other variables, slope is solved on demand as needed and not saved in a grid, since it is not used anywhere else.

$$delX[y][x] = (elevation[y][x+1] - elevation[y][x-1])/2 \qquad (1)$$

$$delY[y][x] = (elevation[y+1][x] - elevation[y-1][x])/2 \qquad (2)$$

$$Slope = \sqrt{delX^2 + delY^2} \qquad (3)$$

All three of the above variables have to be normalized before they can be used in the probability equations.

$$delX_{norm} = \frac{delX}{\sqrt{delX^2 + 1}} \qquad (4)$$

$$delY_{norm} = \frac{delY}{\sqrt{delY^2 + 1}} \qquad (5)$$

$$Slope_{norm} = \frac{Slope}{\sqrt{Slope^2 + 1}} \qquad (6)$$

Calculate Slope Skew. Slope skew is defined as an apparent change in elevation, to reflect conditions such as prevailing sunlight direction, rainfall and wind relative to the direction that a slope is facing. It is not a separate variable passed on to the ecotope modeler, but rather added to the elevation. This is done separately for each ecosystem that is evaluated. It is calculated using the normalized versions of delX and delY as defined in equations (4) and (5). The amount of skew is calculated with the following equation.

$$Skew[i] = (SkewX[i] \times delX_{norm}) + (SkewY[i] \times delY_{norm}) \qquad (7)$$

SkewX[i] and SkewY[i] are defined as the amount of change in altitude for slopes in the x and y directions respectively for each ecosystem.

Calculate Ecotope Weights. For each ecosystem being evaluated, its probability is defined as the product of the individual probabilities in Table 5.
Each ecosystem has a minimum, maximum and smoothing value defined for elevation, relative elevation and slope. The minimum and maximum values define the upper and lower boundaries where the probability is 0.5 The smoothing (S) defines how sharp this boundary is.

Table 5. Pseudo-code showing the calculation of an ecosystems probability

```
CalcEcoWeight()
{
        w_e = (probability due to elevation);
        w_r = (probability due to relative elevation);
        w_s = (probability due to slope);
        // all three are calculated with equations (8, 9)

        return (w_e * w_r * w_s);
}
```

The value that is passed on to the function is first normalized with equation (8). X will be zero for a value exactly in the middle of the defined range, 1.0 at both the upper and lower boundaries, and bigger than 1.0 outside of the defined range.

The probability is calculated using equation (9). It will yield a value between 0.0 and 1.0. Useful values for S (smoothing) ranges from 1 (very smooth crossover) to 10 (sharply defined edge).

$$X = abs((Value - AVS)/Range) \tag{8}$$

$$w = 0.5^{X^s} \tag{9}$$

6.4 Calculate Plant Layout

For each plant type within the ecosystem, a number of plants are generated. This is a function of the density of the plants, and a random offset to ensure enough variation in the representation. For each of these plants, a position is determined by adding random offsets from the center of the tile.

Plants are added to the tile as a position and type. There is no information about the way that it will be rendered. This decision is left to the rendering engine that can choose any appropriate method for display.

6.5 Building the Ground-Cover Texture

All the ecosystems used in the database have a representative ground-cover texture associated with it (See Figure 3). For each of the 17x17 grid-points, the representative ground-cover texture is rendered into the tile's ground-cover texture using a semi transparent mask to blend textures together. Figure 6 show how a number of ground cover textures (the top row in Figure 3) was combined to form a new ground cover texture for a tile.

6.6 Random Values

While a lot of random values are used to generate realistic variation of plants and ecosystems, it is very important to keep the results fully deterministic. This is done by pre-calculating a random lookup table and using a constant offset per tile into the table. The main reason is to ensure consistency of the visual scene. If the camera rotates through 360 degrees, the tile cache will be filled with new tiles, and all the old tiles (including their objects) will be replaced. When the camera looks in the original direction again, the tiles will be recreated and cached. The trees should be in the same position as they were before. By saving the random lookup table as part of the database, it is possible to ensure consistency across a network simulation as well.

Fig. 6. Blending together of ecosystem ground-cover textures to form a new ground cover texture for a tile

7 Results

A program was developed in c++ using DirectX to evaluate the performance of ecotope modeling, and determine its suitability to real-time applications. This section looks at the visual appearance achieved, as well as the speed of the different sections of the algorithm. Five ecosystems where defined as shown in Table 6. All of them are level zero ecosystems. No further splitting of the tiles was done during the test.

Table 6. The five ecosystems used to evaluate the system

Ecosystem	Color			Elevation	Relative elevation	Slope
Dense bush			Min	140	-0.5	0.0
			Max	260	0.1	0.7
			Sharpness	2	1	2
Marshland			Min	-50	-0.5	-0.2
			Max	50	0	0.3
			Sharpness	2	1	2
Small bushes and grass			Min	-50	0.07	-0.2
			Max	350	0.3	0.8
			Sharpness	2	1	2
Grass on steep slopes			Min	-50	0	0.7
			Max	350	1	1.2
			Sharpness	2	1	2
Exposed rock			Min	-50	-1	1.4
			Max	350	1	4.2
			Sharpness	2	1	8

Figure 7 shows both a false color map (on the left) and a color representation (on the right) of the test scene. A single plant has been defined in the dense bush ecosystem.

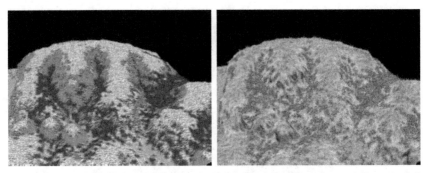

Fig. 7. False color map of the area (left) showing the placement of ecosystems. The color is in accordance with Table 6. On the right, the ground cover textures where replaced with more appropriate textures

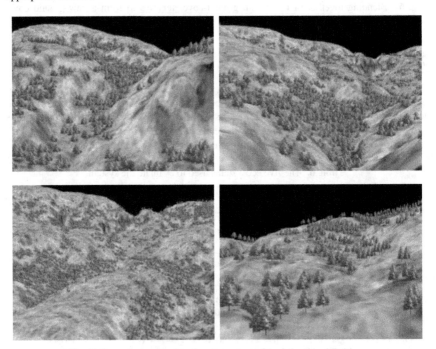

Fig. 8. Four views showing the placement of trees by the algorithm.

Figure 8 show four more views of the terrain. The lack of ecosystems on higher levels can clearly be seen close to the terrain. The trees where rendered as billboards facing towards the camera.

7.1 Performance

The time that it takes to calculate a complete tile is very important. It is possible to have fast simulations with as little as three tile updates per frame. Depending on the exact needs, this allows between one and five milliseconds to calculate a tile. Table 7 shows the average measurements for the different processes. All measurement where made on an Intel PIII with 128 Meg of RAM, and a GeForce graphics card. The only optimizations done was in algorithm design.

Table 7. Time (in ms) to calculate different sections of a tile

Total	4.50 ms
Elevation and relative elevation	0.16 ms
Ecosystem placement and slope	3.60 ms
Calculate plants	0.20 ms
Build textures	0.35 ms

8 Conclusion

Programs like WCS have shown that very realistic images can be obtained from the modeling of ecotopes. I have presented a framework in which these ideas can be simplified to allow the real-time modeling of ecotopes. The advantages of this algorithm is as follows:
1. Compression of natural landscapes. All of the plants are calculated at runtime from a very small description. All the ecosystem information for a complete scene can be described in less than a Meg. This allows the algorithm to run efficiently on machines with limited RAM, freeing up memory resources for other processes.
2. Near real-time execution. Currently the algorithm needs about 15ms per frame to model ecosystems. With optimizations it would be possible to reduce this to as little as 6ms per frame. This is adequate for 30 fps screen updates, and will in the near future (due to faster computers) be fast enough to deliver 60 fps update rates.

References

1. 3D Nature (2000). World Construction Set 5 Users Manual. www.3dnature.com
2. Planetside Software (2000). Terragen Documentation. www.planetside.co.uk
3. Geomantics Ltd (1998). Genesis II Documentation. www.geomantics.com

Seamless Integration of New Control Technologies

Ewout Korpershoek

Topcon Europe B.V., Essebaan 11, 2908 LJ Capelle aan den IJssel, The Netherlands
http://www.topconeurope.com
info@topcon.nl

In the construction industry, the use of survey and positioning technologies is more and more focussing at automation of some of the contractors most valuable assets; heavy machinery.

The use of rotating lasers or sonic devices to take non-contacting reference to a stringline or existing surface, is standard practice on many jobsites today. These systems offer automatic control of the hydraulics of the machine so that the elevation and cross slope of a cutting edge or blade, are set using the laser or sonic reference.

However, systems using laser and/or sonic devices still need reference of manually implanted grade stakes, markers or elevation points. With today's generation of applied survey and positioning technologies, it has become possible to greatly reduce and even eliminate manual stake out work.

Local Positioning Solutions

By directly connecting the digital job design to a tracking total station, the position of the moving machine can be compared continuously to this design. The difference between the two is then fed to the machines' hydraulic controls which automatically control the elevation and slope control of the blade, so that it cuts exactly to the design. The only thing the machine operator needs to worry about is steering his machine, so that he can fully focus on managing his materials.

Benefits

Benefits of such systems are significant reductions of stake out work, no more need for setting up grade stakes & markers every 12 m along a road job, string lines etc.
In addition it also leads to reduction of control checks and re-staking during the process. Furthermore, jobs that used to be complex (non planar jobs like airport run- & taxiways, parking lots with many breaklines, steep vertical curves on landfills or highway exits) can now be done with the same speed and accuracy and ease as a flat job. This results in production increase on virtually every job. Due to the possibility of continuous control, the whole process and flow of activities can be carried out almost uninterrupted and productivity increases are achieved varying from 30% to a 100% for basic road construction, to even 400% for more complex works. And because there is no more human staking and the whole job can now be done with one

C.Y. Westort (Ed.): DEM 2001, LNCS 2181, pp. 112-116, 2001.

survey data set from start to finish, the chance of errors is greatly reduced and leads to a better quality assurance. Another benefit is an increased safety on the jobsite, as no more people are working around the machine for grade checking and manual control of markers and finished surface.

Laser Communication

Specifically for 3D Control of construction machines, Topcon has designed a specific transmitter, that is based on a traditional tracking total station for survey purposes, but utilizes a different principle for control & communication. The GRT-2000 is a tracking total station that continuously tracks the machine during its movements. It has a field computer with the digital design attached to it, and extracts the required slope and elevation data for each point the machine is at. The elevation data is not, like with radio, transmitted to the machine, but a pointed laser beam mounted over the units telescope, sends out a solid laser beam, directly aiming at the right elevation.

A special receiver mounted on the machine's blade, the LS-2000, automatically centers itself to this laser, setting the blade at exactly the right elevation for that particular point. The special sensor is placed at one side of the blade, and a slope sensor which is mounted in the middle of the blade, now takes care of the slope of the other side of the blade. As the total station knows the moving direction of the machine, it can extract the correct required cross slope from the design database. This information is sent to the machine as well, modulated on the same laser beam that's used for controlling the elevation.

By using the laser for direct elevation control and communication, hardly any delay is in the system and high data control rates can be achieved, which is necessary to not limit the machine in its regular working speed, achieve the accuracy required and deliver a smooth end result.

The main reason for this approach is that the servo total station and the moving machine are part of two different physical control loops so that the control system does not suffer from vibrations or unexpected abrupt movements of the machine. This is achieved by tracking the machine only in X and Y directions, the elevation is always controlled by the laser, which is steady at the correct, theoretical height. The smart sensor continuously reads the laser beam and adjusts itself and the blade to the correct height. So in case the machine hits a rock, the 3D receiver will notice immediately and send a correction signal to the machine's hydraulics, without any total station involvement. This is done at 50Hz speed. With this principle, the delay and latency of the system are minimal guaranteeing high accuracy, and smooth end control. A conventional one-man survey system has to track all three position components, and determine positions/corrections for each point in time measured. See Diagram 2 for a graphical comparison of the two methods.

Another important aspect of this solution is the fact that the speed of control is high, and accurate enough to control virtually any machine. Whether it's a fast, steady moving motorgrader, a fast and abrupt moving bulldozer, or a slow asphalt paver. In this way, the components required for 3D control, plug in seamlessly to the same standard machine control system that is used for laser and sonic control.

The GRT-2000 does have all required survey and stake out functionality for both two man as one man use, so if not controlling a machine, it can be used for any regular survey task on site as well.

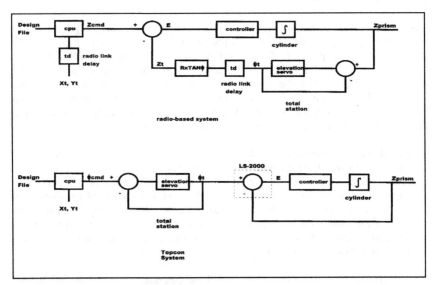

Fig. 1. Control loop principle of GRT-2000 3D Machine Control versus conventional radio based, X,Y,Z controlling total station.

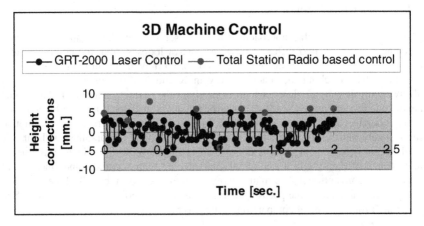

Fig. 2. Comparison of control corrections of Topcon GRT-2000 3D Control system and Topcon AP-L1A tracking total station.

Global Positioning Systems

Where Local Positioning Systems offer accuracies at the mm level, various earthmoving and grading applications are at cm level, and for such applications, the use of GPS becomes extremely viable.

Topcon has developed a 3D-MC™ GPS solution, which is fully integrated with the standard automatic control system on the machine. Using one GPS receiver on the machine's blade to determine elevation, in conjunction with a slope sensor, and an additional touchscreen computer for display and control purpose in the cabin, the operator has fully automatic GPS control. A base station set up over a known reference point provides GPS data.

Whatever the job or application, today's contractor has a choice of state of the art control solutions to best fit his production environment!

Rockingham Motor Speedway Project, Corby UK

Morrison Construction Ltd. From the U.K., and Surfacing Specialists Colas from France, currently have multiple GRT-2000 3D Systems operational at the new to be built 2.4 km Formula 1 circuit in Corby, UK.
In this application, the system not only demonstrates it's high repeatability and accuracy, but also its integrated functionality during all phases of the job.
From the start of the earthworks section, the GRT-2000 was used on a Bulldozer, taking care of site preparation and base layers. Later in the project, a Motorgrader was equipped, for fine grading the track and producing a lime stabilized section. This section was then trimmed using a profiler, also equipped with 3D.
Finally, an asphalt paver produced the asphalt base layers reaching accuracies within the 4 mm range. For the top layers eventually, a sonic averaging system was used to average out the base, and finally result in mm accuracy racing surface for tomorrow's champions.

Zurich Airport Extension Kloten

The family owned Eberhard civil engineering group of Kloten, Switzerland, owns 4 Topcon 3D-MC™ systems. The specification for one of its key projects, an extension to the Zurich airport, is particularly demanding with 24 hours a day working, and high accuracy requirements. The systems are installed on a dozer, two motorgraders and a profiler (trimmer).
The digital job design is directly based on models generated by Siteworks™, a design application running on Bentley's Microstation®, which is directly imported into the

field computer running the 3D system. The direct import saves much time and assures working with the original design data.

After several months of operation, accuracies from 0 to 8 mm are achieved, within 95 percent of the checked points. The main advantage for Mr. U. Koch is the amount of time saved, as stake out work has been virtually eliminated. Another big saving is in material ; with an area covering 550.000 m², and accuracies that are approximately 1 cm tighter then other methods, savings have already exceeded investment in the technology.

Another benefit for the company is the modularity and interchangeability of the systems between the different machines. At Zurich airport, the ability to use a dozer for fine grading is particularly important in coping with weak underground materials. The GRT-2000 system is the only one that offers high speed automatic control of a dozer as if it were a grader.

Author Index

Lecture Notes in Computer Science

For information about Vols. 1–2095
please contact your bookseller or Springer-Verlag

Vol. 2138: R. Freivalds (Ed.), Fundamentals of Computation Theory. Proceedings, 2001. XIII, 542 pages. 2001.

Vol. 2139: J. Kilian (Ed.), Advances in Cryptology – CRYPTO 2001. Proceedings, 2001. XI, 599 pages. 2001.

Vol. 2141: G.S. Brodal, D. Frigioni, A. Marchetti-Spaccamela (Eds.), Algorithm Engineering. Proceedings, 2001. X, 199 pages. 2001.

Vol. 2142: L. Fribourg (Ed.), Computer Science Logic. Proceedings, 2001. XII, 615 pages. 2001.

Vol. 2143: S. Benferhat, P. Besnard (Eds.), Symbolic and Quantitative Approaches to Reasoning with Uncertainty. Proceedings, 2001. XIV, 818 pages. 2001. (Subseries LNAI).

Vol. 2144: T. Margaria, T. Melham (Eds.), Correct Hardware Design and Verification Methods. Proceedings, 2001. XII, 482 pages. 2001.

Vol. 2146: J.H. Silverman (Eds.), Cryptography and Lattices. Proceedings, 2001. VII, 219 pages. 2001.

Vol. 2147: G. Brebner, R. Woods (Eds.), Field-Programmable Logic and Applications. Proceedings, 2001. XV, 665 pages. 2001.

Vol. 2149: O. Gascuel, B.M.E. Moret (Eds.), Algorithms in Bioinformatics. Proceedings, 2001. X, 307 pages. 2001.

Vol. 2150: R. Sakellariou, J. Keane, J. Gurd, L. Freeman (Eds.), Euro-Par 2001 Parallel Processing. Proceedings, 2001. XXX, 943 pages. 2001.

Vol. 2151: A. Caplinskas, J. Eder (Eds.), Advances in Databases and Information Systems. Proceedings, 2001. XIII, 381 pages. 2001.

Vol. 2152: R.J. Boulton, P.B. Jackson (Eds.), Theorem Proving in Higher Order Logics. Proceedings, 2001. X, 395 pages. 2001.

Vol. 2153: A.L. Buchsbaum, J. Snoeyink (Eds.), Algorithm Engineering and Experimentation. Proceedings, 2001. VIII, 231 pages. 2001.

Vol. 2154: K.G. Larsen, M. Nielsen (Eds.), CONCUR 2001 – Concurrency Theory. Proceedings, 2001. XI, 583 pages. 2001.

Vol. 2157: C. Rouveirol, M. Sebag (Eds.), Inductive Logic Programming. Proceedings, 2001. X, 261 pages. 2001. (Subseries LNAI).

Vol. 2158: D. Shepherd, J. Finney, L. Mathy, N. Race (Eds.), Interactive Distributed Multimedia Systems. Proceedings, 2001. XIII, 258 pages. 2001.

Vol. 2159: J. Kelemen, P. Sosík (Eds.), Advances in Artificial Life. Proceedings, 2001. XIX, 724 pages. 2001. (Subseries LNAI).

Vol. 2161: F. Meyer auf der Heide (Ed.), Algorithms – ESA 2001. Proceedings, 2001. XII, 538 pages. 2001.

Vol. 2162: Ç. K. Koç, D. Naccache, C. Paar (Eds.), Cryptographic Hardware and Embedded Systems – CHES 2001. Proceedings, 2001. XIV, 411 pages. 2001.

Vol. 2164: S. Pierre, R. Glitho (Eds.), Mobile Agents for Telecommunication Applications. Proceedings, 2001. XI, 292 pages. 2001.

Vol. 2165: L. de Alfaro, S. Gilmore (Eds.), Process Algebra and Probabilistic Methods. Proceedings, 2001. XII, 217 pages. 2001.

Vol. 2166: V. Matoušek, P. Mautner, R. Mouček, K. Taušer (Eds.), Text, Speech and Dialogue. Proceedings, 2001. XIII, 452 pages. 2001. (Subseries LNAI).

Vol. 2167: L. De Raedt, P. Flach (Eds.), Machine Learning: ECML 2001. Proceedings, 2001. XVII, 618 pages. 2001. (Subseries LNAI).

Vol. 2168: L. De Raedt, A. Siebes (Eds.), Principles of Data Mining and Knowledge Discovery. Proceedings, 2001. XVII, 510 pages. 2001. (Subseries LNAI).

Vol. 2170: S. Palazzo (Ed.), Evolutionary Trends of the Internet. Proceedings, 2001. XIII, 722 pages. 2001.

Vol. 2172: C. Batini, F. Giunchiglia, P. Giorgini, M. Mecella (Eds.), Cooperative Information Systems. Proceedings, 2001. XI, 450 pages. 2001.

Vol. 2174: F. Baader, G. Brewka, T. Eiter (Eds.), KI 2001: Advances in Artificial Intelligence. Proceedings, 2001. XIII, 471 pages. 2001. (Subseries LNAI).

Vol. 2175: F. Esposito (Ed.), AI*IA 2001: Advances in Artificial Intelligence. Proceedings, 2001. XII, 396 pages. 2001. (Subseries LNAI).

Vol. 2176: K.-D. Althoff, R.L. Feldmann, W. Müller (Eds.), Advances in Learning Software Organizations. Proceedings, 2001. XI, 241 pages. 2001.

Vol. 2177: G. Butler, S. Jarzabek (Eds.), Generative and Component-Based Software Engineering. Proceedings, 2001. X, 203 pages. 2001.

Vol. 2180: J. Welch (Ed.), Distributed Computing. Proceedings, 2001. X, 343 pages. 2001.

Vol. 2181: C. Y. Westort (Ed.), Digital Earth Moving. Proceedings, 2001. XII, 117 pages. 2001.

Vol. 2182: M. Klusch, F. Zambonelli (Eds.), Cooperative Information Agents V. Proceedings, 2001. XII, 288 pages. 2001. (Subseries LNAI).

Vol. 2184: M. Tucci (Ed.), Multimedia Databases and Image Communication. Proceedings, 2001. X, 225 pages. 2001.

Vol. 2186: J. Bosch (Ed.), Generative and Component-Based Software Engineering. Proceedings, 2001. VIII, 177 pages. 2001.

Vol. 2187: U. Voges (Ed.), Computer Safety, Reliability and Security. Proceedings, 2001. XVI, 261 pages. 2001.

Vol. 2188: F. Bomarius, S. Komi-Sirviö (Eds.), Product Focused Software Process Improvement. Proceedings, 2001. XI, 382 pages. 2001.

Vol. 2189: F. Hoffmann, D.J. Hand, N. Adams, D. Fisher, G. Guimaraes (Eds.), Advances in Intelligent Data Analysis. Proceedings, 2001. XII, 384 pages. 2001.

Vol. 2190: A. de Antonio, R. Aylett, D. Ballin (Eds.), Intelligent Virtual Agents. Proceedings, 2001. VIII, 245 pages. 2001. (Subseries LNAI).

Vol. 2191: B. Radig, S. Florczyk (Eds.), Pattern Recognition. Proceedings, 2001. XVI, 452 pages. 2001.

Vol. 2193: F. Casati, D. Georgakopoulos, M.-C. Shan (Eds.), Technologies for E-Services. Proceedings, 2001. X, 213 pages. 2001.

Vol. 2196: W. Taha (Ed.), Semantics, Applications, and Implementation of Program Generation. Proceedings, 2001. X, 219 pages. 2001.